THE POETRY OF GEOLOGY

Edited by
Robert M. Hazen

'Poetry is so closely connected with whatever is grand and beautiful, that there is hardly a science or an art which does not possess more or less of it. . . . Shall not geology, which is the first science in affording scope for the imagination, be brought into favor with the Muses, and afford themes for the Poet?'

Edward Hitchcock Jr (1849)

London
GEORGE ALLEN & UNWIN
Boston   Sydney

George Allen & Unwin (Publishers) Ltd,
40 Museum Street, London WC1A 1LU, UK

George Allen & Unwin (Publishers) Ltd,
Park Lane, Hemel Hempstead, Herts HP2 4TE, UK

Allen & Unwin Inc.,
9 Winchester Terrace, Winchester, Mass 01890, USA

George Allen & Unwin Australia Pty Ltd,
8 Napier Street, North Sydney, NSW 2060, Australia

First published in 1982

British Library Cataloguing in Publication Data

Poetry of geology.
   1. Poetry—Themes, motives
I. Hazen, Robert M.
821'.008'36   PR1195.G/
ISBN 0-04-808032-2

Set in 11 on 13 point Bodoni by Bedford Typesetters Limited
and printed in Great Britain by Mackays of Chatham

To Margaret

Preface

Geological poems are among the most entertaining attempts to describe the physical world in which we live. This collection of 'geopoetry,' drawn from British and American sources of the eighteenth and nineteenth centuries, is an outgrowth of bibliographic studies pursued during the past decade. Selections were found in a variety of sources, including religious tracts, medical treatises, geological textbooks and monographs, and periodicals for farmers, miners, educated ladies, and children. The subject matter and tone of these geological poems is no less diverse, with verses ranging from the humorous to the pedantic, advocating various scientific, social, and religious doctrines.

The principal object in assembling this collection has been to provide enjoyment for those who are fascinated by the Earth and its history. This volume is not intended to be an historical analysis of science or poetry, although it is hoped that these selections will demonstrate the widespread popularity and understanding of geology in the nineteenth century. In keeping with the nineteenth-century contents of this book, the format and typography are representative of the small poetry volumes that abounded before 1850. Woodcut illustrations are taken from geological textbooks and treatises of the day. Brief explanatory notes on the source and content of each poem are found at the end of the book.

Librarians at several Washington-area collections provided valuable assistance in locating rare, original editions of British and American poetry. Special thanks are due to the staffs of the United States Geological Survey Library, Reston, Virginia, the National Library of Medicine, Bethesda, Maryland, and the Library of Congress.

R. M. HAZEN
October 1981

Contents

Introduction

Geological phenomena have always captivated mankind. The majesty of giant caverns, the devastation of earthquakes and volcanoes, the incomprehensible antiquity of fossil bones — all are natural wonders that excite the imagination and place man's own stature and history in perspective. At no time were English-speaking people more aware of the wonders of geology than during the first half of the nineteenth century.

The century from 1750 to 1850 produced a revolution in the way man perceived his physical surroundings. The popular enthusiasm which led to geological poetry was in large part an outgrowth of dramatic new attitudes about the age of the Earth, the nature of geological change, and the constancy of living species. In 1750 the Earth was thought by most people to be only a few thousand years old. Mountain ranges, canyons, oceans, and lakes were believed to have been formed by rapid, catastrophic events; the most significant geological occurrence was considered to be the Noachian Deluge. Man's pre-eminent position among a fixed number of immutable species was clearly established by Biblical scholars and natural philosophers. God was the First Cause of everything — of Creation, of the Flood, and the lesser cataclysms.

The scientific rationalism of the eighteenth century led to a search for second causes — the physical means by which God's will is done. The natural laws of mathematics and physics suggested ways in which a universe, once set in motion by the Creator, might continue to operate *without* divine intervention. In the eighteenth century the search for similar laws governing the history and operation of the Earth proceeded in earnest. The Scottish geologist James Hutton (1726–97), for example, concluded that the Earth is inconceivably old, with processes operating today much as in the past. The gradual action of erosion, sedimentation and uplift were recognized as the principal agents of geological change. Mountains and valleys were seen to be the products of thousands of millenia of slow change. Catastrophies — and God's direct intervention in them — were given a minor role in Earth's history.

If Hutton's demonstrations of the Earth's great age and ever-changing surface were radical, then Georges Cuvier's (1769–1832) revelations of animal extinction and ever-changing life forms were even more startling. That God's creatures could cease to exist raised baffling questions about His design in nature and man's own place and permanence in the Cosmos. Cuvier's descriptions and reconstructions of giant extinct mammals at the Museum d'Histoire Naturelle in Paris were followed by dozens of discoveries of fossil bones in both Europe and America. Popular periodicals and newspapers frequently publicized such finds, and the fact of the extinction of life forms was widely accepted by the early nineteenth century.

These three radical concepts — an inconceivably old Earth, evolution of landforms by gradual change, and the extinction of living species — had practical as well as philosophical implications. An understanding of gradual sedimentation and animal extinction led to the recognition that each age of strata has its own unique suite of fossils, which could be used to deduce the geological history of a region. Armed with this knowledge, geologists had mapped much of Britain, western Europe, and eastern North America by the 1820s. Announcements of the discoveries of rich mines, fabulous fossils, and monumental topography appeared almost daily in newspapers and popular periodicals of nineteenth-century Britain and North America. Magazines and textbooks heralded the new geology, and the revolutionary theories of Hutton, Cuvier and others were part of a basic education.

Thus, in a few short decades geology was transformed from a theologically constrained study of catastrophies to a study of vast stretches of time, unimaginable forces, and former worlds of strange fauna and flora. It is little wonder, given this new understanding and the metaphoric power of geological processes, that the science should be allied with another abiding passion of the nineteenth century — poetry.

Poetry was an immensely popular form of entertainment in the first half of the nineteenth century in both Britain and America. Most popular periodicals and many newspapers published poems on a regular basis. Epic book-length poems were read by all educated people, and many of these volumes were as widely known as popular novels of our day. Even scientific textbooks and treatises were commonly enriched by poetry, both as epigrams and within the body of the works.

Poetry was written in response to virtually all human emotions and endeavors. Religion, commerce, the arts, and the sciences all inspired their share of verses. Geology was no exception. Giant fossil bones, fabulous mineral riches, vast caverns, mighty mountains, and fascinating perspectives on time and change were ideal subjects for versification. The poetry of geology was a direct response to the discovery and analysis of these wonders. Edward Hitchcock Jr's essay, 'The poetry of geology,' eloquently expresses the source of inspiration for geological poetry.

The poetry of geology is varied in authorship, style, and content. A few geological poems were contributed by well-known geologists and naturalists, such as George Fleming Richardson (1796–1848) in England and Constantine Rafinesque (1783–1840) in the United States. Another earth scientist, James Gates Percival (1795–1856), was not only one of America's best known poets, but also served as State Geologist in Connecticut and Wisconsin. More commonly, non-geologist poets borrowed themes from geology to enrich their works. James Montgomery (1771–1854) and Felicia Dorothea Hemans (1793–1835) were well-known British poets, whose works, including geologically inspired verses, were reprinted many times on both sides of the Atlantic. The majority of geological poems, however, appear to have been published anonymously in popular literary magazines of the mid-nineteenth century. Nine of the poems in this collection were unsigned.

The subject matter and tone of geological poetry varied greatly. Several poems, including those on Mammoth Cave, earthquakes, and volcanoes, emphasize the overwhelming grandeur of natural phenomena. Other verses, such as 'To a fossil fern' and 'The nautilus and ammonite' focus on smaller scale Earth productions. The vast span of geologic time is a theme of Montgomery's coral island descriptions in *The pelican island* and in Rafinesque's *The former world*.

'Geopoetry' was often written for pure amusement. The whimsical 'Geologist's wife' and 'Epitaph on a mineralogist' were appropriate contributions to the popular literary magazines where they first appeared. The pun-laden verses of John Scafe's *Geological primer* are educational as well as humorous, for many details of British stratigraphy, paleontology, and structural geology are presented in the context of mythical feasts and amours. Several poems in this volume were penned to advocate a position, either religious or social. *Lines made after the great earthquake* was a call for repentence and a return to godly ways. 'The miner lad' and 'The miner's doom' romantically portray the hardships and dangers of mining, and advocate fair conditions and wages for all laborers. A few poems with geologic content are thinly veiled advertisements. 'Meditation on Rhode-Island coal' was an effective endorsement of the anthracite of the Rhode Island Coal Company, and *A poem on the mineral waters of Ballston and Saratoga* served as an elaborate travel brochure for the New York spas.

Whether in whimsy or advocacy, whether about Nature's grandest spectacles or her microscopic forms, the poetry of geology appealed, and still appeals, to man's desire to understand and describe his physical surroundings. As mankind gains new insights about his planet from the vantage of space we must try to preserve this sense of fascination and delight — an aesthetic of science — as we continue to unfold the wonders of geology.

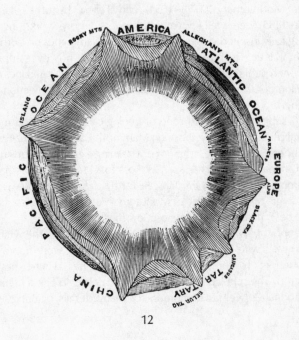

The poetry of geology

An essay by Edward Hitchcock Jr

The poetry of Mathematics, the poetry of Philosophy, and the poetry of Astronomy! How often do we hear these expressions from the real lovers of these sciences, as well as from those who love them only for the sake of their poetry. In fact, we believe that poetry is so closely connected with whatever is grand and beautiful, that there is hardly a science or an art which does not possess more or less of it. Perhaps we ought to say that there is no existence of *anything* in the whole universe without poetry, especially when we use it in its highest sense, and apply it to the divine mind, who created everything with the nicest regard to harmony and order.

And although we may assert with good reason, that poetry is an essential element of all knowledge, yet our perceptions are so feeble, that we are only able to discover it in the most striking examples. Hence, we shall be able to see poetry most readily in the more grand and impressive works of nature, and find those the best Poets, who have the most vivid imaginations, and appreciate every beauty to its fullest extent.

If those thoughts be true, why may we not add Geology to the list of poetical sciences? Why shall not that science, which is the second science in eras and magnitudes, and the first, in affording scope for the imagination, be brought into favor with the Muses and afford themes for the Poet?

Does he ask for something grand and awful? Then I would refer him to those great and powerful changes, which have so essentially altered the appearance of the surface of the globe. Some of these have occurred almost instantaneously; as when a continent or island has, in a very short time, sunk beneath the ocean, or where territory enough for a kingdom, has suddenly arisen into being, from its fathomless depths.

If we wish to employ our imagination in works of Geology, we have but to picture to ourselves a wide fissure in the earth's crust, some hundred miles in length. From its dark depths there suddenly arises and spreads over the surface an amount of melted rock, sufficient to form some of our highest mountains. If we will imagine still farther, let us call up the shades of the *mighty dead*, the Iguanodons, Megatheria, and Dinornithes, and hear them tell their story concerning their former habitations; of the gigantic vegetation and flowers, and of the innumerable variety of animals which swarmed upon the earth in primeval ages, and in their original strength, and then let us see if we have not enough of the wonderful and the marvellous to grace a Poet's lay.

And if we turn our attention to more minute objects, we shall find things even yet more strange. Take the smallest particle of dust that is visible to the naked eye.

The Age of Reptiles.

Here are the remains of animals of the most perfect organization and structure, hundreds of which would find their *world* in a drop of water. Nor are these confined to a single species, but are as infinitely varied as any other class of the animal kingdom.

As poets are ever fond of that which is changeful and varied, let me entreat them to walk with me through some one or more geological periods. Perhaps at our starting point we shall see but little that betokens life, save some pale wrecks or degenerated insects which seem out of their native element, so long have they lived beyond their proper period of existence. Next we come to a lake, or to a river, where we find animal life in the greatest profusion, but of an entirely different character, so that we seem to be in a new country. Passing on through this section, we come upon a vast plain, extending as far as the eye can reach. It is a dreary journey that we have in prospect. The only things to break the monotony, are, now and then, a vestige of some animal which has strayed thus far only to die, or some stalk or leaf, which the winds in their careless sport, have brought hither, or perchance, some *track* of a wandering animal, which time had not completely obliterated.

Now let us suppose our travelers to have become weary of their journey, and to have lain down and slept for *only a few thousand years,* then to have awoke on some bright summer's morn. And what would greet their vision? Instead of the dry and waste sands, they would find themselves in the midst of a rich and luxuriant garden. Above their heads would wave the splendid Magnolias and Tree Ferns, while under their feet would be the softest mosses, and the rarest, richest flowers. Around them the feathered songsters would tune their sweet and melodious lays, and everything would bespeak life, joy, and happiness.

Such fields of wonder and novelty does the Geologist continually travel. And can any one doubt but that they excite the most pleasurable and poetical emotions, in the minds of the contemplators? Can they fail to rouse the imagination in those who are the plainest 'matter of fact' men?

Will not, then, some son of the Muses attempt this task which we have proposed; and, plucking a fresh feather from the bird of Jove, transcribe for mortal eyes and ears, the story of the past, as well as that of the present.

To a fossil fern

Child of an ancient world! o'er whom the storms
   That shatter'd empires silently have roll'd,
   What awful mysteries could'st thou unfold
Of Chance and Change in all their various forms!
Thy frond-like leaves were blooming when in glory,
   Proud Rome and Egypt each beheld its prime,
And doubtless thou could'st tell us many a story
   Of mighty victors of the olden time.
Geology, with microscopic eye,
   Regards thee as a phantom metaphoric;
While Chemistry, whose flight is always high,
   Claims thee as a production meteoric;
But sister Poesy seems half afraid,
And wisely keeps her learning in the shade.

TAB. 129.—THE FLORA OF THE CARBONIFEROUS EPOCH.

(*Designed and drawn by Miss Ellen Maria Mantell.*)

Fig. 1. Auracaria. 2. Asterophyllites comosa. 3. Pandanus. 4. Equisetum. 5. Arborescent fern. 6. Fern. 7. Calamites. 8. Lepidodendron. 9. Sphenopteris.

17

The coal and the diamond

A coal was hid beneath the grate,
    ('Tis often modest merit's fate;)
      'Twas small, and so perhaps forgotten;
Whilst in the room and near of size,
    In a fine basket lined with cotton,
In pomp and state a diamond lies.
    'So, little gentleman in black,'
The brilliant spark in anger cried,
    'I hear, in philosophic clack,
Our families are close allied:
    But know the splendor of my hue,
Excelled by nothing in existence,
    Should teach such little folks as you
To keep a more respectful distance.'

At these reflections on his name,
The coal soon reddened to a flame:
Of his own real use aware,
He only answered with a sneer;
I scorn your taunts, good Bishop Blaze,
    And envy not your charms divine;
For know I boast a double praise,
    As I can *warm* as well as shine.

EXCERPTS FROM
A GEOLOGICAL PRIMER IN VERSE

by John Scafe

Granitogony, or the birth of granite

In ancient time, ere Granite first had birth,
And form'd the solid pavement of the earth,
Stern Silex reign'd, and felt the strong desire
To have a son, the semblance of the sire.
To soft Alumina his court he paid,
But tried in vain to win the gentle maid;
Till to caloric and the spirits of flame
He sued for aid—nor sued for aid in vain:
They warm'd her heart, the bridal couch they spread,
And Felspar was the offspring of their bed:
He on his sparkling front and polished face
Mix'd with his father's strength his mother's grace.
Young Felspar flourish'd, and in early life
With pale Magnesia lived like man and wife.
From this soft union sprang a sprightly dame,
Sparkling with life—and Mica was her name.
Then Silex, Felspar, Mica, dwelt alone,
The triple deities on Terra's throne.
For he, stern Silex, all access denied
To other gods, or other powers beside.
Oft when gay Flora and Pomona strove
To land their stores, their bark he rudely drove
Far from his coast; and in his wrath he swore
They ne'er should land them on his flinty shore.

19

Fired at this harsh refusal, angry Jove,
In terrors clad, descended from above;
His glory and his vengeance he enshrouds,
Involved in tempests and a night of clouds:
O'er Mica's head the livid lightning play'd,
And peals of thunder scared the astonished maid.
To seek her much-loved parents quick she flew;
Her arms elastic round their necks she threw,—
'Thus may I perish, never more to part,
Press'd to my much lov'd sire's and grand-sire's heart!'
So spoke the maid. The thunder-bolt had fled,
And all were number'd with the silent dead.
But interfused and changed to stone, they rise
A mass of Granite towering to the skies.
O'er the whole globe this ponderous mass extends,
Round either pole its mighty arms it bends;
And thus was doom'd to bear in after time
All other rocks of every class and clime.
So sings the bard that Granite first had birth,
And form'd the solid pavement of the earth:
And minor bards may sing, whene'er they list,
Of Argillaceous or Micaceous Schist.

MORAL

Learn hence, ye flinty hearted rocks,
    Your burthens all to bear,
Lest Jove should fix you in the stocks,
    Or toss you in the air.

TO MAKE GRANITE

Of Felspar and Quartz a large quantity take,
Then pepper with Mica, and mix up and bake.
This Granite for common occasions is good;
But, on Saint-days and Sundays, be it understood,
If with bishops and lords in the state room you dine,
Then sprinkle with Topaz, or else Tourmaline.

N.B. The proportion of the ingredients may be varied
ad libitum;—it will keep a long time.

TO MAKE PORPHYRY

Let Silex and Argil be well kneaded down,
Then colour at pleasure, red, grey, green, or brown;
When the paste is all ready, stick in here and there
Small crystals of Felspar, both oblong and square.

TO MAKE PUDDINGSTONE

To vary your dishes, and shun any waste,
Should you have any left of the very same paste,
You may make a plum-pudding; but then do not stint
The quantum of Pebbles—Chert, Jasper, or Flint.

TO MAKE AMYGDALOID

Take a mountain of Wacke, somewhat softish and green,
In which bladder-shaped holes may be every where seen;
Choose a part where these holes are decidedly void all,
Pour Silex in these, to form Agates spheroidal,
And the mass in a trice will be Amygdaloidal.

TO MAKE A GOOD BRECCIA WITH A CALCAREOUS CEMENT

Break your rocks in sharp fragments, preserving the angles;
Of Mica and Quartz you may add a few spangles:
Then let your white batter be well filtered through,
Till the parts stick as firm as if fastened by glue.

TO MAKE A COARSER BRECCIA

For a Breccia more coarse you may vary your matter;
Pound Clay, Quartz, and Iron-stone, moisten'd with water:
Pour these on your fragments, and then wait awhile,
Till the Oxyd of Iron is red as a tile.

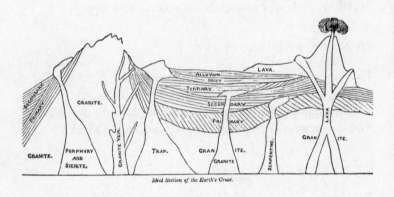

Ideal Section of the Earth's Crust.

When nature was young, and Earth in her prime,
All the Rocks were invited with Neptune to dine.
On his green bed of state he was gracefully seated,
And each as they enter'd was civilly greeted.
But in choosing their seats, some confusion arose,
Much jostling and scrambling, and treading on toes;
Till with some dislocations, and many *wry faces*,
They at length became quiet, and kept their own places.
Reveal, heavenly Muse, for I know thou art able,
How each guest in succession was ranged at the table;
How the dinner was served, and the name of each dish,
Whether Nautilite, Ammonite, tortoise, or fish.

First Granite sat down, and then beckon'd his queen,
But Gneiss stepp'd in rudely, and elbow'd between,
Pushing Mica-slate further; when she with a frown
Cried, 'You crusty, distorted, and hump-back'd old clown!'
But this was all sham,—for to tell you the truth,
They had been the most intimate friends from her youth,
But let scandal cease. See the whole tribe of Slates
All eager and ready to rush to their plates;
Oh heav'ns! how the family pour in by dozens,
Of brothers, and sisters, and nephews, and cousins!
The elder-born Limestones ran in between these,—
They were very well known to be fond of a squeeze.
Now, before we proceed with our story, it meet is
That we hint at th' amours of Calcium and Thetis:

But the tale shall be short. 'Tis agreed by the sages,
Hence sprang all the limestones of different ages:
The oldest look'd white; and no wonder she should,
She had never once dined upon animal food.
Ere these rocks were all seated, the loud sounding call
Of 'Our places! Our places!' rang shrill through the hall.
On hearing the noise, the Muse turn'd round her head,
And saw Porphyry and Eurite—their faces were red.
Then Greenstone and Sienite follow'd behind,—
Their seats were bespoke (they said) time out of mind.
Great Neptune rose up, and then swore in a rage
That each rock should be seated according to age;
'But let those (where the register cannot be found
Either under the water or on the dry ground)
Not presume to take regular seats at the table,
But change places with others, whenever they're able.'
Thus the last-mentioned rocks were obliged to retire,
Though their ages were book'd in the office of fire:
(This they said,) but no soul would go there to inquire.
Leaning over old Gneiss and the Slate-rocks they stood,
Or else press'd between them, whenever they could.
Gay Serpentine, clad in a livery of green,
At Mica-slate's feet during dinner was seen;
Among the first class it was publicly said,
He had often been found fast asleep in her bed.

When these rocks were thus settled, and quiet restored,
The others more orderly march'd to the board.
Say, Muse, who is he that is just walking in?
O! his name is as harsh and as rough as his skin,
He's a cousin of Slate, but he looks wild and cracky,
And is known as the far-famed illustrious Grau-Wacce.
Younger Slate-rocks, with Sand-stone, then came side by
    side,
And he, the Great Limestone, of limestones the pride,
Who has caves with wild echos resounding and vocal,
And is called by the masons *grey marble entrochal*.
The next were a grave-looking set on the whole,
Who came in a group to accompany Coal.
Coarse grit-stones, with sand-stones, and clay-binds, and
    shale,
Some were hard, some were soft, some were dingy, some
    pale;
They oft proved deceitful when thought very sound,
For they had many *faults*, which they hid under ground.
Red Sand-stone came after, and licking his lips,
He brought in the salt, on a salver of Gyps.
To two sister limestones he had a strong bias,
The one was Magnesian, the other was Lias.
Though the former look'd sallow, he press'd the dear
    charmer
So close, his attentions did sometimes alarm her:
But Lias was *flat*, and seem'd sombre and dull,
For with shell-fish and lizards her stomach was full.
Then Oolites, with sandstones, and sand red and green
In a crowd, near the top of the table were seen.
The last that were seated were Chalk-marl and Chalk,
They were placed close to Neptune, to keep him in talk.

Now the God gave his orders, 'If more guests should come,
Let them dine with the Lakes, in a separate room.
As for Gravels, and Black-earth, and other gross livers,
They may feast out of doors, by the sides of the rivers.
Kill Aurochs and Mammoths, not heeding their groans,
But let them take care of the teeth and the bones.'
The strata from Paris arrived very late,
With letters, requesting a chair and a plate.
'*Eh bien*,' said the God, with a good natured air,
'*Faites entrer Monsieur le Calcaire Grossier;*
Let him and his friends at a sideboard be placed,
And with Cerites and Lymnites the covers be graced.'

Now, Muse, raise thy voice, and be kind to our wishes,
And tell us the names of the principal dishes.
To Chalk, preserved palates and fossil Echini
Were handed in Clam-shells more pearly than China.
Then Alcyonites, Nautilites, graced a tureen,
With Belemnites tastefully stuck in between.
The Oolites were served with a wonderous profusion
Of Bivalves, dished up in apparent confusion.
There Trigonias, Anomias, and Arcas were placed,
And each rock took the species that tickled his taste.
At this juncture some Limpets were sent in on one dish,
From our worthy friend Halifax, vicar of Standish.
Now oviparous creatures, in which back-bone is,
Were hash'd with remains of the Cornua Ammonis.
They were bringing in more; but great Neptune cried
      'Halt!
Place no vertebral animal lower than Salt:

Those grits, and those shales, hold inflammable matter,
Let no lizards or fishes e'er smoke on their platter:
Give them fern-leaves, and palm-stalks, and such like
  spare diet,
And Coal and Pyrites will keep very quiet.
The Great Limestone full plates of Encrini will want,
To some of the others a few you may grant:
Feed the lower with Coral; and some of the Slates
May have Shell-fish most sparingly spread on their plates.
The eldest born Limestone, whose colour so white is,
Of Mica and Talc-slate the well known delight is;
With Granite and all the old Rocks she shall fare,
And dine on bright Crystals, both costly and rare.'
The commands of great Neptune were duly observed,
And their dinner in state was most splendidly served.
Yellow Topaz, red Garnets, and Emeralds, we're told,
Were sent under covers of bright burnish'd gold,
With Schorl's red and green, and blue Sapphires and Beryl;
But the Muse thought this diet too arid and steril,
So she moved from the seat of such infinite splendour;
For, like us, she loved something more juicy and tender.
Long lasted the dinner—No rock from his seat
Ever moved, or evinced the least wish to retreat;
And old Neptune found out, as the wise ones aver,
When the rocks are once seated, they love not to stir,
So he rose unobserved, and began to retire;
But 'tis whispered the Sea-God already smelt fire.
Be this as it may—a deep hollow sound,
Still nearer and nearer was heard under ground;

'Twas the chariot of Pluto,—in whirlwinds of flame
Through a rent in the earth to the dinner he came.
'Oh, by Styx and by Hecate, my rage I wont smother,
What—Nep gave a feast, without asking his brother?
Though I am King of Hell—what, am I such a sinner
That I can't be invited to smoke after dinner?
Let Nep with his waves and his waters all go to—
I'll make the rocks dance, or my name is not Pluto.'
Thrice he stamp'd in a rage, and with crashes like thunder
The earth open'd wide, and the rocks burst asunder,
And the red streaming lava flow'd over and under.
It spread far and wide, till grim Pluto said 'Halt!'
And ranged it in columns and files of Basalt!
For he saw Neptune coming, collecting his might,
And roaring and raising his waves for the fight.

Now were Eurite and greenstone beginning to *run*,
Which Hutton and Hall said was excellent fun.
But a rock-rending scene in the sequel it proved,
E'en the hard heart of Porphyry was melted and moved.
And many a rock the muse could not draw nigh to,
She saw very plainly was softened *in situ*.
Now thick vapours of Sulphur and clouds black as night,
Roll'd in volumes, and hid the whole scene from the sight;
And the Muse told the Poet 'twas time to take flight:
Adding this—'My good fellow, pray leave off your writing,
We have had quite enough both of feasting and fighting.'

On the entrance to the Mammoth Cave

From the Pen of a Tourist

To what new world are we now tending?
  Is it one of pain and sin?
Whither are our footsteps wending?
  How the darkness hems us in!
That stream of water falling quick—
  How ominous its tone:
We pause—we tremble—we are sick,
  And feel deserted—lone.
The water falleth in a pit,
  Like tears of those who weep;
Far over head deep gloom doth sit,
  And Nature is asleep.
On, on we go—soon cheer'd in soul,
  For we are told that light,
Bursting from stern night's control,
  Is near to trance our sight.

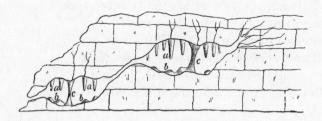

Mammoth Cave

by George D. Prentice

All day, as day is reckoned on the earth,
I've wandered in these dim and awful aisles,
Shut from the blue and breezy dome of heaven,
While thoughts, wild, drear, and shadowy, have swept
Across my awe-struck soul, like spectres o'er
The wizard's magic glass, or thunder clouds
O'er the blue waters of the deep. And now
I'll sit me down upon yon broken rock,
To muse upon the strange and solemn things
Of this mysterious realm.
                  All day my steps
Have been amid the beautiful, the wild,
The gloomy, the terrific. Crystal founts,
Almost invisible in their serene
And pure transparency—high, pillar'd domes
With stars and flowers all fretted like the halls
Of Oriental monarchs—rivers dark
And drear and voiceless as oblivion's stream,
That flows through Death's dim vale of silence—
All fathomless, down which the loosened rock
Plunges until its far-off echoes come
Fainter and fainter like the dying roll
Of thunders in the distance—Stygian pools
Whose agitated waves give back a sound
Hollow and dismal, like the sullen roar

In the volcano's depths—these, these have left
Their spell upon me, and their memories
Have passed into my spirit, and are now
Blent with my being till they seem a part
Of my own immortality.

                  God's hand,
At the creation, hollowed out this vast
Domain of darkness, where nor herb nor flower
E'er sprang amid the sands, nor dews nor rains
Nor blessed sunbeams fell with freshening power,
Nor gentle breeze its Eden-message told
Amid the dreadful gloom. Six thousand years
Swept o'er the earth ere human foot-prints marked
This subterranean desert. Centuries
Like shadows came and passed, and not a sound
Was in this realm, save when at intervals,
In the long lapse of ages, some huge mass
Of overhanging rock fell thundering down.

Its echoes sounding through these corridors
A moment, and then dying in a hush
Of silence such as brooded o'er the earth
When earth was chaos. The great mastodon,
The dreaded monster of the elder world,
Passed o'er this mighty cavern, and his tread
Bent the old forest oaks like fragile reeds,
And made earth tremble.—Armies in their pride
Perchance have met above it in the shock
Of war, with shout and groan and clarion blast,
And the hoarse echoes of the thunder gun;
The storm, the whirlwind and the hurricane
Have roared above it, and the bursting cloud
Sent down its red and crashing thunder-bolt;
Earthquakes have trampled o'er it in their wrath,
Rocking earth's surface as the storm-wind rocks
The old Atlantic;—yet no sound of these
E'er came down to the everlasting depths
Of these dark solitudes.
                              How oft we gaze
With awe or admiration on the new
And unfamiliar, but pass coldly by
The lovlier and the mightier! Wonderful
Is this lone world of darkness and of gloom,
But far more wonderful yon outer world
Lit by the glorius sun. These arches swell
Sublime in lone and dim magnificence.
But how sublimer God's blue canopy

Beleaguered with his burning cherubim
Keeping their watch eternal! Beautiful
Are all the thousand snow-white gems that lie
In these mysterious chambers gleaming out
Amid the melancholy gloom—and wild
These rocky hills and cliffs, and gulfs—but far
More beautiful and wild the things that greet
The wanderer in our world of light—the stars
Floating on high like islands of the blest—
The autumn sunsets glowing like the gate
Of far-off Paradise—the gorgeous clouds
On which the glories of the earth and sky
Meet and commingle—earth's unnumbered flowers
All turning up their gentle eyes to heaven—
The birds, with bright wings glancing in the sun,
Filling the air with rainbow miniatures—
The green old forests surging in the gale—
The everlasting mountains on whose peaks
The setting sun burns like an altar-flame—
And ocean, like a pure heart rendering back
Heaven's perfect image, or in his wild wrath
Heaving and tossing like the stormy breast
Of a chained giant in his agony.

Lines made after the great earthquake, in 1755

WHICH SHOOK NORTH AND SOUTH AMERICA,
WITH GREAT DESTRUCTION IN CALES, IN LISBON,
AND MOST OF THE ADJACENT KINGDOMS

Awake New-England now and view
Thy God is just—his words are true
His outstretch'd arms will rule us by,
That power which sinner's deeds defy.

In seventeen hundred fifty-five
When vice its empire did revive
Consuming fire, a jealous God
Call'd on New-England with his rod.

The rod God's voice bid earth to shake,
Tremors which caus'd our hearts to ache,
Did cause confusion in our minds
As may be seen in following lines.

The moon arose all fair and bright
And shed forth silver rays that night
We in our beds had found repose
But soon from them we trembling rose.

Hark New-England—Loud the thunder
Awake ye sinners from your slumber,
By deeds no longer do defy
That power descending from the sky.

We through our habitations past,
Fearing our souls in hell be cast
While shattering buildings loudly cry'd
Your deeds have God's great power defy'd.

These judgments he has sent to you
Know he is just—his ways are true.
If you will not with him comply
With fiercer judgements he will try.

America I loudly call,
Return to God as one and all.
Tis mocking God to make delay,
Come seek for mercy, never stay.

While hapless Cales and Lisbon shake
From God their judgement did partake
His mercy fav'd his grace adore
Which spar'd New-England's happy shore.

While Lisbon's sands roll as the waves
And thousands cast into their graves,
While Korah and his sons are lost
His power secures and guards our coast.

Warned in time to him repair,
His mercies still abounding are,
Then you shall know the grace and bliss.
Of Jesus and his righteousness.

Thus thence in heaven you'll rest from care,
From sin, from woes, from sorrows, where
In fields of endless light, you'll rise
In mansions far beyond the skies.

In mansions there in bliss unknown,
High in salvation near the throne,
Where saints and angels all shall join
In songs and anthems all divine.

Sensations and reflections

by Flaccus

I lay at morn half-conscious of the dawn:
My pausing soul, touched by returning sense
Of duty, yet unwilling to forbear
Her rosy journey through the land of dreams,
Hung doubtful like a cloud 'twixt heaven and earth
Midway, or like a failing bird that long
Had beat the ether of sublimer spheres,
Reluctant downward drooped; when suddenly
Shouted a mighty voice, and truant Reason
Leaped to her post: deep inward groans, as though
The uttered grief of Earth's capacious breast,
Came up, and her profound and solid frame
Shuddered beneath me, that my lifted couch
Quivered unsteady as a floating bark:
Wonder and awe oppressed me, and I felt
Held for the instant in the hand of God!
I knew the frantic EARTHQUAKE in his car
Had rattled by, and laughed; and visions swift
Trooped o'er my brain, of horrors manifold
That have befallen when this mighty orb
Cracked like a globe of glass, alarming nations
With the wild thunder; whose deep-rung vibrations
Ran jarring from the tropic to the pole:

When cities shook, unseated; and loose walls,
And staggering towers across the peopled streets,
Nodded and knocked their heads in ponderous ruin
Deep-burying all below: wildest convulsion
Of all that agitate the frame of Nature!

How solemn 't is upon the rocking deep
To feel the mastery of the lawless waves!
Helpless, uncertain but their treacherous arms
That lift us up so high may part apace,
And down to dark and unimagined horrors
Leave us to sink: what double terror then
When sober Earth mimics the reeling sea!
And plains, upheaving into billows, yield
Unsolid to the foot of man and beast;
When our sure dwelling, like a foundering bark,
Pitches and rolls, the plaything of those strange
Unnatural waves, while hideous underneath
Yawn greedier caves than deepest ocean hides,
Glutted with fragments of the shipwrecked earth,
Clashing and plunging down! O! let us kneel
And offer up the incense of our thanks
To Him that spared us blow so horrible,
And only laid his lightest finger-touch
(Gently as though the frozen frame of Earth
Had barely shivered with the wintry chill,
Or as some wing of passing angel, bound
From sphere to sphere, had brushed the golden chain
That hangs our planet to the throne of God,)

To jog our sluggish memories that His hand
Upholds, commands us still.
                              Tremble, ye rich!
Where were your mansions now, had He, indignant,
Pushed from their firm foundations? Where your lands,
Had His unpitying hand, withdrawing, left
Their unsupported burthen to go down
To the strange bottom of some new-born sea?
Tremble, ye great! ye puny apes of power,
That with mock-majesty misrule the earth,
Where were ye now, had His scorned sceptre
In earnest ire fell on your heads? Ye! whom
This lightest pulse of the almighty heart
Quails to your just dimensions! Yet wherefore
Bid warning to the rich, the great, alone,
When ALL should reverent bow: have we not all
A stake more priceless than command or gold—
His favor? Let our thousand hearts, that stirred
Like leaves at this hushed whisper of His might,
Pause, and with inward probing seek the cause.
That drew the chiding of the sovereign down.
Are his commands forgot?—our solemn duties
Ill-done? or left, through folly's vain pursuit,
Untouched? Then let us wisely take new heart,
And from the couch that trembled at HIS touch
Rise up, resolved to bend us to our task
With manly zeal, that at the close of day
We may go up to meet our Master's face,
And claim the promised wages without shame!

Thus lulled to calm reliance in the fold
Of 'everlasting arms,' should lurking tempests
Spring sudden upon sleeping Nature; should
Rebellious fires, that in embowelled Earth
Lie prisoned; rise, and writhing to be free,
Burst her centripetal and iron bands—
Unhinging continents, uprooting mountains,
Until her ragged quarters all at large
Fly diverse into space, leaving a gap
Of yawning night, wherein our helpless form
Drops like a stone, piercing an unknown gulf,
Too deep for thought to sound—how would we smile
At baffled Fate! safe in the precious trust
That we had won us an Almighty friend,
And he would lend us wings to break our fall!

EXCERPTS FROM
THE PELICAN ISLAND

by James Montgomery

This was the landscape stretch'd beneath the flood:
—Rocks, branching out like chains of Alpine Mountains;
Gulfs intervening, sandy wildernesses,
Forests of growth enormous, caverns, shoals;
Fountains upspringing, hot and cold, and fresh
And bitter, as on land: volcanic fires
Fiercely outflashing from earth's central heart,
Nor soon extinguish'd by the rush of waters
Down the rent crater to the unknown abyss
Of Nature's laboratory, where she hides
Her deeds from every eye except her Maker's:
—Such were the scenes which ocean open'd to me;
Mysterious regions, the recluse abode
Of unapproachable inhabitants,
That dwelt in everlasting darkness there.
Unheard by them the roaring of the wind,
The elastic motion of the wave unfelt;
Still life was theirs, well pleasing to themselves,
Nor yet unuseful, as my song shall show.

Here, on a stony eminence, that stood,
Girt with inferior ridges, at the point,
Where light and darkness meet in spectral gloom,
Midway between the height and depth of ocean,
Mark'd a whirlpool in perpetual play,

As though the mountain were itself alive,
And catching prey on every side, with feelers
Countless as sunbeams, slight as gossamer:
Ere long transfigured, each fine film became
An independent creature, self-employ'd
Yet but an agent in one common work,
The sum of all their individual labours.
Shapeless they seem'd, but endless shapes assumed;
Elongated like worms, they writhed and shrunk
Their tortuous bodies to grotesque dimensions;
Compress'd like wedges, radiated like stars,
Branching like sea-weed, whirl'd in dazzling rings;
Subtle and variable as flickering flames,
Sight could not trace their evanescent changes,
Nor comprehend their motions, till minute
And curious observations caught the clew
To this live labyrinth,—where every one,
By instinct taught, perform'd its little task;
—To build its dwelling and its sepulchre.
From its own essence exquisitely modell'd;
There breed, and die, and leave a progeny,
Still multiplied beyond the reach of numbers,
To frame new cells and tombs; then breed and die
As all their ancestors had done,—and rest,
Hermetically seal'd, each in its shrine,
A statue in this temple of oblivion!
Millions of millions thus, from age to age,
With simplest skill, and toil unweariable,

No moment and no movement unimproved,
Laid line on line, on terrace terrace spread,
To swell the heightening, brightening gradual mound,
By marvellous structure climbing tow'rds the day.
Each wrought alone, yet all together wrought,
Unconscious, not unworthy, instruments,
By which a hand invisible was rearing
A new creation in the secret deep.
Omnipotence wrought in them, with them, by them;
Hence what Omnipotence alone could do
Worms did. I saw the living pile ascend,
The mausoleum of its architects,
Still dying upwards as their labours closed:
Slime the material, but the slime was turn'd
To adamant, by their petrific touch;
Frail were their frames, ephemeral their lives,
Their masonry imperishable. All
Life's needful functions, food, exertion, rest,
By nice economy of Providence
Were overruled to carry on the process,
Which out of water brought forth solid rock.

Atom by atom, thus the burthen grew,
Even like an infant in the womb, till Time
Deliver'd ocean of that monstrous birth,
—A coral island, stretching east and west,
In God's own language to its parents saying,
'Thus far, nor farther, shalt thou go; and here
Shall thy proud waves be stay'd:'—A point at first
It peer'd above those waves; a point so small,
I just perceived it, fix'd where all was floating;
And when a bubble cross'd it, the blue film
Expanded like a sky above the speck;
That speck became a hand-breadth; day and night
It spread, accumulated, and ere long
Presented to my view a dazzling plain,
White as the moon amid the sapphire sea;
Bare at low water, and as still as death,
But when the tide came gurgling o'er the surface,
'Twas like a resurrection of the dead:
From graves innumerable, punctures fine
In the close coral, capillary swarms
Of reptiles, horrent as Medusa's snakes,
Cover'd the bald-pate reef; then all was life,
And indefatigable industry;
The artizans were twisting to and fro,
In idle-seeming convolutions; yet
They never vanish'd with the ebbing surge,
Till pellicle on pellicle, and layer
On layer, was added to the growing mass.

Ere long the reef o'ertopt the spring-flood's height,
And mock'd the billows when they leapt upon it,
Unable to maintain their slippery hold,
And falling down in foam-wreaths round its verge.
Steep were the flanks, sharp precipices,
Descending to their base in ocean-gloom.
Chasms few, and narrow, and irregular,
Form'd harbours, safe at once and perilous,—
Safe for defence, but perilous to enter.
A sea-lake shone amidst the fossil isle,
Reflecting in a ring its cliffs and caverns,
With heaven itself seen like a lake below.

Compared with this amazing edifice,
Raised by the weakest creatures in existence,
What are the works of intellectual man?
Towers, temples, palaces, and sepulchres;
Ideal images in sculptured forms,
Thoughts hewn in columns, or in domes expanded,
Fancies through every maze of beauty shown;
Pride, gratitude, affection turn'd to marble,
In honour of the living or the dead;
What are they?—fine-wrought miniatures of art,
Too exquisite to bear the weight of dew,
Which every morn lets fall in pearls upon them,
Till all their pomp sinks down in mouldering relics,
Yet in their ruin lovelier than their prime!

—Dust in the balance, atoms in the gale,
Compared with these achievements in the deep,
Were all the monuments of olden time,
In days when there were giants on the earth:
—Babel's stupendous folly, though it aim'd
To scale heaven's battlements, was but a toy,
The plaything of the world in infancy:—
The ramparts, towers, and gates of Babylon,
Built for eternity,—though where they stood,
Ruin itself stands still for lack of work,
And Desolation keeps unbroken sabbath;—
Great Babylon, in its full moon of empire,
Even when its 'head of gold' was smitten off,
And from a monarch changed into a brute;—
Great Babylon was like a wreath of sand,
Left by one tide, and cancell'd by the next:—
Egypt's dread wonders, still defying Time,
Where cities have been crumbled into sand,
Scatter'd by winds beyond the Libyan desert,
Or melted down into the mud of Nile,
And cast in tillage o'er the corn-sown fields,
Where Memphis flourish'd, and the Pharaohs reign'd;—
Egypt's gray piles of hieroglyphic grandeur,
That have survived the language which they speak,
Preserving its dead emblems to the eye,
Yet hiding from the mind what these reveal;

—Her pyramids would be mere pinnacles,
Her giant statues, wrought from rocks of granite,
But puny ornaments for such a pile
As this stupendous mound of catacombs,
Fill'd with dry mummies of the builder-worms.

EXCERPT FROM
GREENLAND

by James Montgomery

From eve till morn strange meteors streak the pole;
At cloudless noon mysterious thunders roll,
As if below both shore and ocean hurl'd
From deep convulsions of the nether world.
Anon the river, boiling from its bed,
Shall leap its bounds and o'er the lowlands spread,
Then waste in exhalation,—leaving void
As its own channel, utterly destroy'd,
Fields, gardens, dwellings, churches and their graves,
All wreck'd or disappearing with the waves.
The fugitives that 'scape this instant death
Inhale slow pestilence with every breath;
Mephitic steams from Schapta's smouldering breast
With livid horror shall the air infest;

And day shall glare so foully on the sight,
Darkness were refuge from the curse of light.
Lo! far among the glaciers, wrapt in gloom,
The red precursors of approaching doom,
Scatter'd and solitary founts of fire,
Unlock'd by hands invisible, aspire;
Ere long more rapidly than eye can count,
Above, beneath, they multiply, they mount,
Converge, condense,—a crimson phalanx form,
And rage aloft in one unbounded storm;
From heaven's red roof the fierce reflections throw
A sea of fluctuating light below.
—Now the whole army of destroyers, fleet
As whirlwinds, terrible as lightnings, meet;
The mountains melt like wax along their course,
When downward, pouring with resistless force,
Through the void channel where the river roll'd,
To ocean's verge their flaming march they hold;
While blocks of ice, and crags of granite rent,
Half-fluid ore, and rugged minerals blent,
Float on the gulph, till molten or immersed,
Or in explosive thunderbolts dispersed.
Thus shall the Schapta, towering on the brink
Of unknown jeopardy, in ruin sink;
And this wild paroxysm of frenzy past,
At her own work shall Nature stand aghast.

Look on this desolation:—mark yon brow,
Once adamant, a cone of ashes now:
Here rivers swampt; there valleys levell'd, plains
O'erwhelm'd;—one black-red wilderness remains,
One crust of lava, through whose cinder-heat
The pulse of buried streams is felt to beat;

These from the frequent fissures, eddying white,
Sublimed to vapour, issue forth like light
Amidst the sulphury fumes, that drear and dun
Poison the atmosphere and blind the sun.
Above, as if the sky had felt the stroke
Of that volcano, and consumed to smoke,
One cloud appears in heaven, and one alone,
Hung round the dark horizon's craggy zone,
Forming at once the vast encircling wall,
And the dense roof of some Tartarean hall,
Propt by a thousand pillars, huge and strange,
Fantastic forms that every moment change,
As hissing, surging from the floor beneath,
Volumes of steam the imprison'd waters breathe.

Epitaph on a mineralogist

by Felicia Dorothea Hemans

Stop, passenger, a wondrous tale to list—
Here lies a famous mineralogist!
Famous, indeed,—such traces of his power
He's left from Penmanbach to Penmanmawer,—
Such caves, and chasms and fissures in the rocks,
His works resemble those of earthquake shocks;
And future ages very much may wonder
What mighty giant rent the hills asunder;
Or whether Lucifer himself had ne'er
Gone with his crew, to play at foot-ball there.
His fossils, flints and spars of every hue
With him, good reader, here lie buried too!
Sweet specimens, which toiling to obtain,
He split huge cliffs like so much wood in twain:
We knew, so great the fuss he made about them,
Alive or dead, he ne'er would rest without them,
So to secure soft slumber to his bones,
We paved his grave with all his favorite stones.
His much loved hammer's resting by his side,
Each hand contains a shell-fish petrified;
His mouth a piece of pudding stone encloses,
And at his feet a lump of coal reposes:
Sure he was born beneath some lucky planet,
His very coffin plate is made of granite!

Weep not, good reader! He is truly blest,
Amidst chalcedony and quartz to rest—
Weep not for him! but envied be his doom,
Whose tomb, though small, for all he loved had room
And, O ye rocks! schist, gneiss, whate'er ye be,
Ye varied strata, names too hard for me,
Sing 'O be joyful!' for your direst foe,
By death's fell hammer, is at length laid low.
Ne'er on your spoils again shall ——— ——— riot,
Shut up your cloudy brows, and rest in quiet!
He sleeps—no longer planning hostile actions,—
As cold as any of his petrifactions;
Enshrined in specimens of every hue,
Too tranquil e'en to dream, ye rocks, of you.

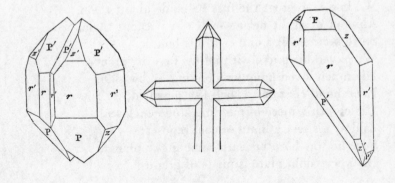

The miner lad

Nay, don't despise the Miner-lad,
  Who burrows like the mole;
Buried alive, from morn to night,
  To delve for household coal—
Nay, miner-lad, ne'er blush for it,
Though black thy face be, as the pit!

As honorable thy calling is
  As that of hero lords,
They owe to the poor Miner-lad
  The ore that steels their swords—
And perils, too, as fierce as theirs
In limb and life, the Miner shares!

Ye gayest of the gaudy world,
  In gold and silver bright,
Who, but the humble Miner-lad,
  Your jewels brought to light?
Where would be your gold and silver,
But for yonder delver?

Ye brows of pearly diadems,
  Who sit on lofty thrones,
Smile gently on the Miner-lad
  Who wrought your precious stones.
And rescued from their iron bond
The ruby and the diamond!

Ye instruments of brass, that pierce
　　The ear with trumpet sound,
Your notes, but for the Miner-lad,
　　Had slumbered under ground—
Nor imaged bronze, nor brazen gate,
Had graced the trophies of the great!

Then don't refuse the Miner-lad
　　The crust of bread—his prayer!
Beneath that blackest face of his
　　He hides a heart as fair!
The toil of his bare brawny arm
All, all our hearts and houses warm!

The miner's doom

Written for the *London Mining Journal*,
by the Author of the 'Syne Exile's Return'

'Twas evening, and a sweeter balm on earth was never
   shed,
The sun lay in his gorgeous pomp on ocean's heaving bed,
The sky was clad in bright array, too beautiful to last,
For night, like envy, scowling came, and all the scene
   o'ercast.

'Tis thus with hope—'tis thus with life, when sunny
   dreams appear,
The infant leaves the cradle-couch to slumber on a bier;
The rainbow of our cherish'd love, we see in beauty's eye,
That glows with all its mingled hues, alas! to fade and die!

'Tis dark, still night, the sultry air scarce moves a leaf
   or flower;
The aspen, trembling, fears to stir, in such a silent hour;
The footsteps of the timid hare, distinctly may be heard
Between the pauses of the song of night's portentious
   bird,—
And in so drear a moment, plods the miner to his toil,
Compelled refreshing sleep to leave, for labor's hardest
   moil;
By fate's rude hand, the dream of peace is broken and
   destroyed—
The savage beast his rest can take, but man must be
   denied!

And why this sacrifice of rest?—did not the Maker plan
The darksome hours for gentle sleep, the day for work by
   man?

Yes!—but the mighty gods of earth are wiser in their
laws—
They hold themselves with pride to be their Creator's first
great cause.

The miner hath his work begun, and busy strokes resound,
Warm drops of sweat are falling fast—the Coal lies piled
around.
And what a sight of slavery!—in narrow seams compressed
Are seen the prostrate forms of men to hew on back and
breast,
Fainting with heat, with dust begrimed, their meagre
faces see,
By glimmering lamps that serve to show their looks of
misery.
And oft the hard swollen hand is raised to wipe the
forehead dews;
He breathes a sigh for labor's close, and then his toil
renews.

And manly hearts are throbbing there—and visions in
that mind
Float o'er the young and sanguine soul, like stars that
rain and shine.
Amid the dreariness that dwells within the cavern's gloom
Age looks for youth to solace him—waits for his fruits to
bloom.
Behold! there is a careless face bent from yon cabined
nook;
Hope you may read in his bright eye—there's future in
his look;
Oh, blight not, then, the fairy flower, 'tis heartless to
destroy
The only pleasure mortals know—anticipated joy!

Oh, God! what flickering flame is this?—see, see again its
  glare!
Dancing around the wiry lamp, like meteors of the air.
Away, away?—the shaft, the shaft!—the blazing fire flies;
Confusion!—speed!—the lava stream the lightning's wing
  defies!
The shaft!—the shaft!—down on the ground, and let the
  demon ride
Like the sirocco on the blast—volcanoes in their pride!
The choke-damp angel slaughters all—he spares no living
  soul!
He smites them with sulphureous brand—he blackens
  them like coal!

The young—the hopeful, happy young—fall with the
  old and gray,
And oh, great God! a dreadful doom, thus buried to decay
Beneath the green and flowery soil whereon their friends
  remain—
Disfigured, and perchance, alive—their cries unheard and
  vain?
Oh, Desolation! thou art now a tyrant on thy throne,
Thou smilest with sardonic lip to hear the shriek and
  groan!
To see each mangled, writhing corse to raining eyes
  displayed—
For hopeless widows now lament, and orphans wail
  dismayed.

Behold thy work! The maid is there, her lover to deplore;
The mother wails her only child, that she shall see no more;
An idiot sister laughs and sings—oh, melancholy joy!—
While bending o'er her brother dead, she opes the
  sightless eye.

Apart, an aged man appears, like some sage David oak,
Shedding his tears, like leaves that fall beneath the
woodman's stroke;
His poor old heart is rent in twain—he stands and weeps
alone—
The sole supporter of his house, the last, the best is gone!

This is thy work, fell tyrant!—this the miner's common lot!
In danger's darkling den he toils, and dies lamented not.
The army hath its pensioners—the sons of ocean rest,
When battle's crimson flag is furled, on bounty's downy
breast;
But who regards the mining slave, that for his country's
wealth
Resigns his sleep, his pleasures, home, his freedom and
his health?
From the glad skies and fragrant fields he cheerfully
descends,
And eats his bread in stenchy caves, where his existence
ends.

Aye, this is he whom masters grind, and level with the
dust—
The slave that barters life, to gain the pittance of a crust.
Go, read your pillard calendar, the record that will tell
How many victims of the mine in yonder churchyard
dwell.
Hath honor's laurels ever wreathed the despot's haughty
brow?—
Hath pity's hallowed gems appeared when he in death
lay low?
Unhonored is his memory, despised his worthless name—
Who wields in life the iron rod, in death no tear can claim!

EXCERPTS FROM
VISITS TO THE MANTELLIAN MUSEUM

by George F. Richardson

The Mantellian Museum

'Tis indeed a world of wonder,
Found within the earth and under;
Fancied forms and wild chimeras,
Creatures of primeval aeras,
Startling all our ancient notions,
Showing lands of old were oceans;
Showing oceans once were dry,
As the mountains old and high!
Wondrous shapes, and tales terrific,
Told in Nature's hieroglyphic;
Written in her countless volumes,
Graven on her granite columns!
Showing many a strangest mystery,
From her ancient, wondrous history.
Forms as wild as fancy wishes,
Monster lizards, stony fishes;
Fragments of the lost amphibia,
Here a femur,—there a tibia;—
Here the monster mammoth sleeping,
There the giant lizard creeping,—
Beings of a tropic nature,
Crocodile and alligator;
Fragments vast of lost creations,
Relics of earth's first formations;

57

Here the snake, the lizard there,
With the tiger and the bear!
Monsters from beneath the waves,
With the creatures hid in caves,
Brought in later days to light,
From their dens of stalagmite!

Yet these giant forms tremendous,
Creatures wondrous, wild, stupendous,—
Huge,—that fancy cannot frame them;
Wild,—that language may not name them,
Differing from a world like this,
Each and all were framed for bliss;
Form'd to share, without alloy,
Each its element of joy,
By that Power that rules to bless,
All were made for happiness!

Nautilus.

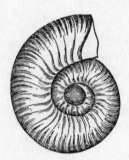

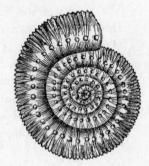

Ammonites.

The nautilus and the ammonite

The Nautilus and the Ammonite,
　　Were launch'd in friendly strife;
Each sent to float, in its tiny boat,
　　On the wide wild sea of life!

For each could swim on the ocean's brim,
　　And when wearied its sail could furl;
And sink to sleep in the great sea deep,
　　In its palace all of pearl!

And theirs was a bliss, more fair than this,
　　That we feel in our colder time;
For they were rife, in a tropic life,
　　In a brighter, and better clime!

They swam 'mid isles whose summer smiles
　　No wintry winds annoy:
Whose groves are palm—whose air is balm—
　　Where life is only joy!

They sailed all day through creek and bay,
　　And traversed the ocean deep;
And at night they sank on a coral bank,
　　In its fairy bowers to sleep!

And the monsters vast of ages past,
    They beheld in their ocean caves;
They saw them ride in their power and pride,
    And sink in their deep sea graves!

And hand in hand, from strand to strand,
    They sailed in mirth and glee;
These fairy shells, with their crystal cells,
    Twin creatures of the sea!

And they came at last, to a sea long past,
    But as they reached its shore,
The Almighty's breath spoke out in death,
    And the Ammonite lived no more!

And the Nautilus now, in its shelly prow,
    As over the deep it strays;
Still seems to seek, in bay and creek,
    Its companion of other days!

And thus do we, in life's stormy sea,
    As from shore to shore we roam,
While tempest-tost, seek the loved, the lost,
    But find them on earth no more!

Yet the hope how sweet, again to meet,
    As we look to a distant strand;
Where heart finds heart, and no more they part,
    Who meet in that better land!

*A ryghte trewe storie of a waulke and taulke
abowte geologye ande historye*

God prosper longe our Ladye Queene,
   Our menne of scyence alle!
What pleasynge waulkes, what learnedde taulkes,
   On Sussex Downes befalle!

Mantell, whoe late toe Lewes broughte
   His followers, fonde and trewe;
Now clymbed the Steyning hilles, and soughte
   'Fresh fyeldes and pastures newe!'

And showed againe, o'er vale and hille,
   With learned taulke and toyle,
The deedes of olde, and older stille,
   The wonders of the soyle!

Fyrst, att the ryver halted wee,
   Whyle Mantell toke his stande,
And tolde the marvelles of the sea,
   And changes of the lande!

'The insecte smalle,' quod he, 'the whyle
   Itt flytts among the flowres,
Thinkes them eternall: do ye smyle?
   Itts errour is but owres!

'Wee, tooe, throughoute lyfe's lyttell daye,
   Looke owre eache tranquil scene,
And fondlie thinke 'twill be for aye,
   And soe hath ever bene!

'But knowe, thatt once no ryver flowed
  Throughoute these smyling fyeldes;
Butt farre off waters drayned the landes,
  And rann thro' distant wealdes!

'And whenn some vaste expansyve force
  Broke upp the ocean's bedde,
'Twas thenn this ryver founde itts cowrse,
  And thro' these valleys spredde!

'And soe, when wee shall vanyshed bee,
  Like change shall then come owre;
The sea be lande, the fyelde a strande,
  The rivere flowe noe more!'

And now o'er hylle, and mounte, and dale,
  His followers Mantell broughte;
And whyle he told the varyed tale,
  This was the lore he taughte:

'The distant wealdes ye gaze upon,
  Once swarmedde with monsters rare;
There ranged the vast Iguanodon,
  The Hylaeosaurus there!

'And later yet a sea owrspredde
  The spot where nowe wee waulke;
And this was once an ocean's bedde,
  The ocean of the chaulke!

'And seas more late, in forme and date,
  Spredde owre the self-same strande;
And manye a change, most wylde and strange,
  Reversed the sea and lande.

'And later stylle, o'er yonder hylle,
  Didde tropyeke creatures roame:
The wild horse, deere, founde pasture here,
  The elephaunte a home!'

And thus, owre valley and owre mounte,
  Didde Mantell holde hys cowrse;
And pawsing laste beside a founte,
  He there described its sowrce.

'This stone of sand, on which I stande,'
  He sayde, the stream besyde,
'Beares deepe and darke the rypple marke,
  Worne by a ryver's tyde.'

Agayne hee tolde the storie olde,
  Yett ever, ever newe,
Of changes wyde, in lande and tyde,
  That earthe and oceane knewe!

'Butt I will cease, and holde mye peace,'
  Enthusyaste Mantell saydde,
'Whyle cleare and bryghte before youre syghte
  The charmes of Nature spredde.

'For, harke! from hylle and vale so stylle
  Ascends her evenyng hymne,
That nowe dothe rayse her Maker's prayse,
  And breathes alle love toe Hym!

'And marke her fyeldes, her woodes, her wealdes,
  Her panorama vaste;
And see the whyle the sunne dothe smyle
  Hys bryghtest and his laste!'

For joyes most sweete are alsoe fleete,
  Now twylyghte's shadowes felle;
Night threwe owre alle her spangledde palle,
  And Mantell badde—Farewelle!

Nowe ye who blame this verse, so lame,
  Writt by unlearnedde elfe,
Thynke not hys lore, as myne, was poore,
  But goe next tyme yourselfe!

You'll synge, I ween, 'Long lyve owr Queene,
  And Mantell, long lyve hee;
And whenn hee waulkes, and when hee taulkes,
  Maye I bee there to see!'

A meditation on Rhode-Island coal

I sat beside the glowing grate, fresh heaped
    With Newport coal, and as the flame grew bright—
The many-coloured flame—and played and leaped,
    I thought of rainbows and the northern light,
Moore's Lalla Rookh, the Treasury Report,
And other brilliant matters of the sort.

And last I thought of that fair isle which sent
    The mineral fuel. On a summer day
I saw it once, with heat and travel spent,
    And scratched by dwarf oaks in the hollow way;
Now dragged through sand, now jolted over stone—
A rugged road through rugged Tiverton.

And hotter grew the air, and hollower grew
    The deep-worn path, and horror-struck, I thought,
Where will this dreary passage lead me to?—
    This long, dull road, so narrow, deep, and hot?
I looked to see it dive in earth outright;
I looked—but saw a far more welcome sight.

Like a soft mist upon the evening shore,
    At once a lovely isle before me lay;
Smooth, and with tender verdure covered o'er,
    As if just risen from its calm inland bay;
Sloped each way gently to the grassy edge,
And the small waves that dallied with the sedge.

The barley was just reaped—its heavy sheaves
　　Lay on the stubble field—the tall maize stood
Dark in its summer growth, and shook its leaves—
　　And bright the sunlight played on the young wood—
For fifty years ago, the old men say,
The Briton hewed their ancient groves away.

I saw where fountains freshened the green land,
　　And where the pleasant road, from door to door,
With rows of cherry trees on either hand,
　　Went wandering all that fertile region o'er—
Rogue's Island once—but, when the rogues were dead,
Rhode Island was the name it took instead.

Beautiful island! then it only seemed
　　A lovely stranger—it has grown a friend.
I gazed on its smooth slopes, but never dreamed
　　How soon that bright beneficent isle would send
The treasures of its womb across the sea,
To warm a poet's room and boil his tea.

Dark anthracite! that reddenest on my hearth,
　　Thou in those island mines didst slumber long,
But now thou art come forth to move the earth,
　　And put to shame the men that mean thee wrong;
Thou shalt be coals of fire to those that hate thee,
And warm the shins of all that under-rate thee.

Yea, they did wrong thee foully—they who mocked
    Thy honest face, and said thou wouldst not burn;
Of hewing thee to chimney-pieces talked,
    And grew profane—and swore, in bitter scorn,
That men might to thy inner caves retire,
And there, unsinged, abide the day of fire.

Yet is thy greatness nigh. I pause to state,
    That I too have seen greatness—even I—
Shook hands with Adams—stared at La Fayette,
    When barehead in the hot noon of July,
He would not let the umbrella be held o'er him,
For which three cheers burst from the mob before him.

And I have seen—not many months ago—
    An eastern governor, in chapeau bras
And military coat, a glorious show!
    Ride forth to visit the reviews, and ah,
How oft he smiled and bowed to Jonathan!
How many hands were shook, and votes were won!

'Twas a great governor—thou too shalt be
    Great in thy turn—and wide shall spread thy fame,
And swiftly—farthest Maine shall hear of thee,
    And cold New-Brunswick gladden at thy name,
And faintly through its sleets, the weeping isle
That sends the Boston folks their cod, shall smile.

For thou shalt forge vast rail-ways, and shall heat
  The hissing rivers into steam, and drive
Huge masses from thy mines, on iron feet,
  Walking their steady way, as if alive,
Northward, till everlasting ice besets thee,
And south as far as the grim Spaniard lets thee.

Thou shalt make mighty engines swim the sea,
  Like its own monsters—boats that for a guinea
Will take a man to Havre—and shalt be
  The moving soul of many a spinning jenny,
And ply thy shuttles, till a bard can wear
As good a suit of broadcloth as the mayor.

Then we will laugh at the winter when we hear
  The grim old churl about our dwellings rave:
Thou from that 'ruler of the inverted year,'
  Shalt pluck the knotty sceptre Cowper gave,
And pull him from his sledge, and drag him in,
And melt the icicles from off his chin.

Heat will be cheap—a small consideration
  Will put one in a way to raise his punch,
Set lemon-trees, and have a cane plantation—
  'Twill be a pretty saving to the *Lunch*.
Then the West India negroes may go play
The banjo, and keep endless holiday.

EXCERPTS FROM
A POEM ON THE MINERAL WATERS
OF BALLSTON AND SARATOGA

by Reuben Sears

Land of my birth! where first the vital air
Of heaven I drew, and first mine opening eyes
Beheld the world's fair frame; whose fields and groves,
In childhood and in youth, my feet have trod;
Thy name is known abroad, and sweetly join'd
With health and pleasure, joy and gay delight.

What draws them here? what but the precious gift
Of bounteous heaven, that in this favor'd spot,
Hath caused to spring waters of life, that heal
The maladies of man, and cheer his heart.

Thy glory lies within thy vales, that wind
Their course along thy centre, and abound
With Mineral Springs, that inexhaustible hold forth
The cup of health and joy, to all that come.

Here nature so hath form'd the soil, so laid
And so arrang'd the Min'ral substances,
Earths, stones and ores, that the sweet veins of water
Coursing the well prepared ground, imbibe
The richest qualities, and issue forth
In never failing founts, to bless mankind.

What, and by what process form'd, these waters are,
Let learned Chymists tell; from whence derived
That fine etherial spirit, that pervades
The agitated mass, and to it gives
Pungency of taste and quick'ning power;
From what source proceeds the strengthening iron,
The salt cathartic, and each different kind
Of rich ingredient contain'd therein.

Skilful Geologists may search the vales,
The streams, and hills surrounding, and disclose
The various strata, that compose the ground,
The fine silicious sand, and stiff blue clay,
The schist argillous, and the lime-stone rock,
And whate'er else may serve to form a soil,
Of such peculiar nature, to produce
Fountains of such unrival'd excellence.

As if suspended o'er eternal fire,
From sight conceal'd, these fountains boil and toss,
In restless agitation, evermore.
Up from the bottom comes the rushing gas,
Hast'ning from its imprison'd state to burst
Into the open regions of the air.
Hence the continual tossing of the founts;
A lively spirit hence through ev'ry part,
Pungent and exhilarating is diffus'd.

But soon the evanescent spirit flies,
If from the parent fount you separate
A portion of the water, and expose
To atmospheric air. The subtle gas
Quickly escapes, and leaves a stagnant mass
Vapid, saline, and loathsome to the taste.

Clear and transparent are these precious founts,
As purest water of the pebbled brook.
No dull opaque their chrystal clearness dims,
Nor floating mote their purity impairs.
Not George's sacred lake, frequented erst,
By superstition's children, to obtain
For holy water its pellucid wave,
Presents a fairer mirror to the eye.

Clear as they are, these waters yet contain
The elements of grosser substances,
Held in solution by the powerful hand
Of the carbonic gas. Here float unseen,
As chymical analysis hath shown,
Iron and magnesia, salt and lime;
Which, with th' enliv'ning gas, the fluid give
Virtues medicinal, removing oft
Divers diseases, that infest the frame
Of frail and mortal man. Redundant bile,
And ev'ry gross secretion, that obstructs
The nice form'd channels of the human frame,
And checks the stream of life, are hence expell'd.

The vital flood flows free, and quick, and pure;
The languid nerves are strung with tension new;
Disorder'd stomachs rectified, and health,
Vigour, and sprightfulness are felt again.

Come to these fountains then, ye sons of sloth!
Pamper'd with luxury, and bloated full
With those gross humors, from which active toil
And plainer fare preserve the lab'ring class,
Who spurn the fell, inebriating bowl:
Drink of these waters, and throw off the load,
That bears on nature with oppressive weight,
And for consuming fever food supplies.

Ye! who in crowded cities live immur'd,
Midst dust, and smoke, and exhalations foul,
From mingled masses of corruption drawn;
Where nought but frost can purify the air,
And temp'rate months alone can be enjoy'd;
Now while the summer's heat oppressive reigns,
Augmented by reflection, and the breeze,
Obstructed by the close built town, fans not
With undulations free your sultry dwellings,
Come to these rural seats, where the sweet air
Of purest heaven you'll breathe, where unconfin'd
The cooling breezes play, and from th' effects
Of nerve relaxing heat these Springs supply
A kind restorative, not known elsewhere.

The call is heard. From ev'ry part arrive
Th' afflicted children of disease and pain,
To try the virtues of these healing founts.
Nor this alone. The wealthy and the gay,
Forth from the cities and from southern climes,
Flock to these Springs, what time the glorious sun
Reigns in full power upon his northern throne.
Nought then is seen but crowded carriages,
And thronging Visitants. Crowd succeeds crowd,
In quick succession, like the restless waves.
From Boston's eastern shore to Georgia's clime
Far distant in the south, and e'en the Isles
Of western India, here the strangers come.
The polish'd multitudes fill up and throng
Our little towns, and o'er these rural scenes,
Splendor, and life, and gaiety diffuse.

Hence in these vales, at places where the Springs
Break from the earth, two thriving villages
Have risen, Ballston and Saratoga.
Where not long since extensive woods prevail'd
And dreary solitudes, by savage men
Inhabited alone and beasts of prey,
Now rise to view the seats of polish'd life;

Well-peopled villages, in which are seen
The neat, convenient dwelling, and the store
Fill'd with the products of far distant climes,
The sacred spire ascending into heaven,
That calls to prayer and praise a Christian people,
The shop mechanic, and the school-house, where,
The little swarming tribes are duly taught,
The trav'ller's inn, in which, when faint and weary,
Refreshment, rest, and comfort he may find;
Hotels of large extent, expressly form'd
The vast concourse of strangers to receive,
Commodious, pleasant, serv'd in highest style,
Where wealth and leisure find a choice retreat,
And fashion sports her gay, bewitching charms.

These vales, so lately wild, have thus become
The gay resort of fashion and of wealth,
Disease's hope, and leisure's sweet retreat,
And such, no doubt, will evermore remain,
While these rich fountains boil, and men delight,
From crowded cities and oppressive heat
Flying, to quaff the cool delicious draught,
Where smiling nature all around invites,
And free and pure, ambrosial breezes play.

The geologist's wife

Adieu then, my dear, to the Highlands you go,
Geology calls you, you must not say no:
Alone in your absence I cannot but mourn,
And yet it were selfish to wish your return.

No, come not until you have searched through the gneiss,
And marked all the smoothings produced by the ice;
O'er granite-filled chinks felt Huttonian joy,
And measured the parallel roads of Glenroy.

Yet still, as from mountain to mountain you stride,
In visions I'll walk like a shade by your side;
Your bag and your hammer I'll carry with glee,
And climb the raised beaches, my own love, with thee.

Me, too, you'll remember, for love claims no less,
And all your proceedings a fondness confess;
Each level you take, be it not from the sea,
But above the dear place where your Susan may be.

Let everything mind you of tender relations—
See, even the hard rocks have *their* inclinations!
Oh, let me believe that, wherever you roam,
The axis of *yours* can be nowhere but—home!

Suppose that you find on the mountains of Lorn,
A boulder that long since from Nevis was torn.
'T will seem like that fond one who left his own shore,
'Perhaps to return to Lochaber no more.'

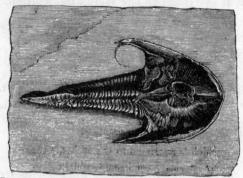

TAB. 136.—CEPHALAPSIS LYELLII: FROM GLAMMIS, IN FORFAR-
SHIRE; A FISH PECULIAR TO THE OLD RED SANDSTONE SYSTEM.
(*Discovered by Mr. Lyell.*)

And if, in your wanderings, you chance to be led
To Ross-shire or Moray, to see the Old Red,
Oh still, as its mail-covered fishes you view,
Remember the colour is love's proper hue.

Such being your feelings, I'll care not although
You're gone from my side—for a fortnight or so;
But know, if much longer you leave me alone,
You may find, coming back you have two wives of stone!

SELECTIONS FROM
THE WORLD, OR INSTABILITY

by C. S. Rafinesque

Mineral springs

The fountains of the earth are earthy pores,
The sweat and moisture of this globe exuding.
How various and unsteady in their sizes,
Contents and functions? Few are always pure,
But liquid fluids of many kinds they throw,
Sweet or impure, both cold and tepid, warm
Or hot; that gently rise, or bubbling boil,
Nay spout on high. Now nearly dry becoming,
Or full their basins filling to the brim.
Not only water flows from earthly springs,
But mineral fluids, holding sulphur, iron,
Acids and gases, lime, and many salts.
Naphtha and oils from fountains seldom flow;
Yet there are such, even liquid pitch
In bubbles bursting underground, in lakes
Expanding; thro' volcanic regions, prone
To offer firy springs, in heat evolving:
While spungy ground, or marshy soil conceal
Of lurid swamps the deadly hues and mire.
Where none arise, where liquid outlets scarce,
Or if the soil they shun, a desert dry
The earth becomes; and if no fluid could moisten
This globe, it would have been a dreary wild,
Unfit for life, where life should be extinct.

Volcanoes

See where the ground in trembling fever quakes
And darts galvanic fires; the clouds of smoke
Ascend on high, the bolts to heaven fly
In all directions; Ashes fall like snow,
And scorch the ground; the burning lava boils,
Like melted iron flows, and desolation
Is spreading far: high hills arise, where none
Before had stood, while others fall or sink.
The fields of men, their homes, their cattle, towns,
And cities proud are swept away by turns.
Upon the earth in various places, high
Or low, arise the hills or lofty cones,
Which bear within their hollow bowels, hot
And awful fires that rocks and metals burn.
Thro' one or many mouths their dingy smokes
Evolving, dreadful loud explosions follow,
To warn and frighten man. In full eruption
The mountain roars and blazes lurid flames.
Showers of ashes, gravel, fill the sky,
And far away to distant regions fly.
The burning lava soon overflows the brim,
In streams of fire upon the sides expands,
To desolate and spoil the blooming ground,
A soil fertile with glowing rocks to fill.

View of Mount Ætna.

Of such volcanoes, dreadful blazing mountains,
I dared to reach the brim, and throw my eyes
Thro' clouds of smoke into their boiling fires.
An awful sight, that makes the stoutest heart
To quiver, wonder, and exclaim, how great
The works of God! But he has will'd these throes
And dismal fires to cool the heated earth,
And warn mankind, that they depend on him.
While from this very power, follows good,
The ashy rains of dust and gravel hard,
Soon crumble and become a fruitful soil,
Where thrive the olive and the vine, of peace
And joy the emblems; overlooking all
The dangers, man there silently admires,
The power that from evil can evolve
A greater good, and fertilize the soil:
While earthly heat is thrown into the air,
To lessen central fires and cool the globe.

Chapter five. The former Earth

CATACLYSMS, FLOODS AND FOSSILS

But while surveying thus the actual earth,
Her changeful scenes; the times recall to mind
Of other ancient changes, ruinous traces,
With memories of cataclysms: events
Of yore by us recorded or surmised,
Which thro' the maze of time we search and find.
When lofty minds delight to raise awhile
The gloomy veil of time and ages past,
Beyond Memory's hold, and Clio's reach
They search unwritten pages, words unspoken,
Medals engraved by Nature's potent hand.
They soar throughout the skies, and ask the suns
When born? how made? and scattered thro' space,
To light and warm the planets, comets, moons.
How rolling worlds were thrown to wheel around,
In splendid homes prepar'd, adorn'd for men
And beings numberless, since born therein.
They sink beneath the soil to seek below
Within the deepest graves records of life;
Their epitaphs of time, reveal, explain.
Of nations sunk to dust almost unknown,
Through various languages no longer spoken,
Through crumbling monuments and relics faint,
They trace the steps and deeds, their arts unfold.
Within the earthly bowels in rocky tombs
They find the bones and shells of buried bodies,

Or woody fragments, formerly partaking,
Enjoying life. Their existence revealed,
A useful lesson teaches; the law of change
Fully confirms, without exceptions ruling
The flying Orbs, and moving living beings.
Meantime in these, and ev'ry where, we may
The mighty hand of God, perceive, adore.

By flood or floods, by many revolutions,
By Cataclysms, successive changes felt,
We may account for rocky tombs involving
These relics, once a softer bed presenting.
A great deluge, a mighty flood of waters,
Come once to overwhelm the earthy globe
From whence we hardly know, yet often dare
Vainly surmise. Some say a comet flew
Too near; while others think a change of motion
Accounts for it; or shock of many fluids
In either case produc'd. We may ascribe
It, if we like, to sinking, or upheaving
Of continents, land regions widely changed;
The ocean swell'd to mountain tides of woe,
Abyss of water spreading desolation
By breaking thro' the solid earthy crust.
Whatever was the cause, by mighty spell,
In overflowing waves, the soil was drowned;
And overwhelming all that stood before
Their way, with rainy floods combining to
Destroy the human race, the animals

And plants that liv'd upon the earth: except
The few that were in mountains sav'd, in arks
Or places of refuge, escaping death
Together with the swimming water tribes.
    But searching minds have lately proclaimed
The awful theme of many floods of yore,
And clysmian deeds of partial causal scope,
Anterior and posterior to the last
Greater deluge; that have destroy'd the lives
Of many beings, even wat'ry tribes:
Ere men had come to share this earthly home.
Two cataclysms at least have since been felt
By men, dismay and horror scattering
Afar. The last was equally severe,
And split the land into the actual fragments.

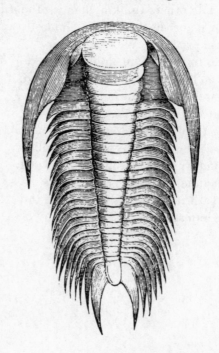

Sunk in the strata of hard rocky stones,
Or beds of slate and sand, are shells and fishes
Once dwelling far within the ocean deep;
But now removed to mountains high and steep,
By sunken sea or lifted land. Besides
The bones of animals and plants, that move
Or grow upon the land: yet now entombed
Not far apart, upon each other met
In superpos'd position, often mingled.
There deeply buried as in their last grave
They have become the medals of this globe;
The evidence of successive creations,
Of living forms now chang'd or quite unknown,
By HIM who never ceases giving life;
Who said, let there be life, and they were born.

If medals struck by nations, cities, kings,
Reveal, recall their names, their deeds and dates
These fossil medals struck by life and death,
Reveal the forms, the existence, sad fate
Of countless beings; names receiving now
From us, when brought to light from their dark tombs.
Some of their deeds also may be imprest
Upon their frames, localities and shapes;
But bear no dates, except the local signs
That successive convulsions indicate,
And we restore, comparing sites of graves.

Even their modes of death, or how extinct,
Is oft obscure, or liable to doubts;
If overwhelm'd by water, fire or mud,
A flood, a stream, a current strong and wide,
Eruptions of volcanoes, rising tides,
Or any other awful kind of fate.
To trace the time of each destructive power,
Respective ages ascertain and fix,
Is arduous task beyond the human ken.
But ev'ry thing by daring man is tried,
And floods of many kinds were thus invented,
Suppos'd, in order to account for each
Stratum of fossil relics in decay
Entomb'd, or faint impressions left in stones
By living stamps destroy'd. The vain surmises
Of such prolific floods, of wide extent
And baneful nature, are not always true,
Never were universal on our globe.
But many local floods have taken place
And yet occur; some fatal, desolating
A tract of country, overwhelming towns
With men and cattle: others even less
In cruelty, are only seen to spread
Over a small extent, and fewer lives
Destroy, of living beings and rooted plants.
Yet both these cataclysms unequal means
Display; the elements are all, employ'd
To wage a war against mankind and life.

The air and winds; the waters, waves and streams;
Earthquakes and spouts, tornadoes, storms and floods:
Eruptive matter, thunderbolts and fires.
They all combine to awe the human race,
By turns assail the earth, and dire effects,
Confin'd to narrow limits, oft produce.
Under the sea, the ocean bitter waves,
Volcanic deep eruptions rage also,
That scatter death among the finny tribes:
The waves themselves to boiling heat reduce.

To ask, when born these fossils were? is idle,
Nay worse, unwise; the countless ages of
Their existence can only be surmised
By guess, comparing depths of graves and sites.

To ask how born, and why no longer now?
Is bold. How can we hope to know, detect,
The ways of God in active mood employed?
Some may contend that many times his power
Upon the earth was felt to bid new life:
But others deem that once alone exerted
To ev'ry thing gave life, by single act
Creating wonders wise beyond belief,
That by successive change unfold themselves.
The earth herself is thought alive, and all
Within; which living power can endow
The very stones with life: they crystalize
In forms quite regular with lines and sides,
In straight or curved angles sharp or flat.

If by increase of action modified
Successive sportive forms arise, combine,
In changeful moods to frame, produce, become
All that we see, with organs, life, endowed;
'Tis but of God the active power still,
Thro' laws of wisdom, change, exerting skill.

From crystals bright and gems so fair and pure
Of atoms form'd in series superposed,
To vegetating cells and tubes minute,
That in combining, vessels, fibres, wood
Become, disclosing art and wise design;
Growing by fluids circulating up
And down, from roots to stems convey'd in plants
Or Trees, inward a latent motion having
Obtain'd. From these but fewer changes may
Produce the motions of spontaneous fixt
Polyps and animals; next spring at last
The moving beings, freely ranging far:
Whose moulds were cast by will divine and wise.
Each growing from their original germs,
As plants from buds and seeds, while stony gems
From molecules arise: and altogether
In elements the stream of life imbibe.

But who shall dare to scan the hidden course
Of this process divine that bids to be?
And all is born to live in changeful mood,
So slowly newer shapes assuming, that
By mortal eyes, but seldom 'tis perceived.

What is an age to God? or thousand years?
Hardly a day, an hour, or even less.
He bids all things to be, and they appear.
He chose they should forever change, and this
They do by human eyes unseen, because
Only awhile we live. Yet men and cattle,
The dogs and beasts, and all the trees or plants,
That we have kept for ages under view
Or cultivation, have in many ways
Their colors, shapes and fruits so often changed,
That this process the dullest sight may strike,
And can't escape a keen investigation.
From this we may presume the same to happen
To other things and bodies, slower still
Or quite beyond the human reach and notice.
But when, and how, and why? are questions bold:
Let wiser minds resolve and answer, when
Longer experience, the truth may teach.
   I will not say with him, Lamark, who dreamt
Of late upon this curious subject, that
This spreading globe, with all its boasted ruins,
Was once a ball of water filled with life,
And atoms quite minute, by heat and light
Of life endow'd; who moving, mixing, changing,
Growing and dying to decay, and sink,
Out of organic ashes, made whatever
We see on land, and all the solid bodies
Inert or living, stones and rocks and mountains,
As well as plants and moving animals.

This theory so fanciful, has few
Believers or supporters; yet we find
That many deem the limy rocks by shells
Alone once made, and others will ascribe
To trees the birth of fossil coals; because
Forsooth, they hold some shells and wood entomb'd.
Graves were not built of human bones, although
Many as yet they hold conceal'd inside.

There is no strange conceit upon this score,
Or any other subject of proud lore,
That has not been by learned men supposed
Or vainly dreamt, to scan, explain and tell
The why of ev'ry thing. When plausible
Hypotheses are built in harmless fancy,
They are mere curious themes of no importance.
But when they ground their visions strange and wild
Upon belief at variance with facts
Or truth, in order to support the creeds
Dogmas or tenets held: they cease to be
Mere harmless dreams, and weapons may become
Of angry strife. Whoever seeks with care
The real truth, of such ought to beware:
And never bow the head to absurd thoughts,
Nor worship learned idols, seldom trustful,
Who worse than idols made by human hands,
In baneful mental bondage keep the mind.

In caves, plaster, clay, and other soils
Are found the bones of beasts so strange and huge,
As stagger human faith in times of yore.
Formerly thought the bones of giants, such
They were declar'd by learned wonders seekers:

Until in later times Cuvier, was born
Whose lofty mind the truth surmising said,
As if a Deity; *arise again*
To view, you beings of the earliest days!
He took their bones and set them side to side,
Until their former frames became restored:
A kind of resurrection taking place,
By skulls and teeth with joints and claws united.
These skeletons were made to stand upright
As when alive, and show the framing structure
Of bodies in decay restor'd to view.

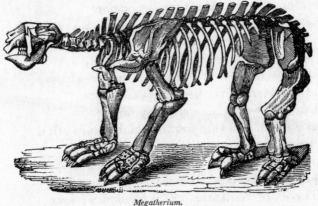

Megatherium.

When once in any science the path is open,
The lesser minds can follow on the steps
Of daring pioneers: thus yearly are
Now brought to light, the fragments of the tombs,
Where living tribes met their early fate.
Th' enquiring mind in this another theme
Has found, to think upon or dream awhile.
When were these beings born and ceased to live?

The why and how? are now the questions, which
Cuvier himself has hardly dar'd unfold:
But bolder minds have tried to make their lives
Agree with strange opinions and beliefs.

   Belief is never proof, conviction flows
From holy truth: but truth by diff'rent minds
Conceiv'd, appears in various shades and forms,
That give belief to some, but certainty
To few: Nor proofs to ev'ry mind convey.
That there has been upon this earthly globe
Another race of living beings, born
To dwell and roam, to sport and feed, as we
Now do, is truth. Also that long before
They dwelt on land, and the dry soil appeared
To be their home; there was another breed
Of water beings swarming in the waves,
Of polyps, shells and crabs, with fishes, whales,
And monsters of the deep: In early ages
When yet the ocean over many lands
Was spread, and this youthful globe was bath'd
In briny tears, or healthy dews and fluids,
Forming around the whole a liquid veil,
Where islands stood as many spots apart.

   These are the truths, but if beyond we soar
And seek minute details, or to explain
Every thing we see, in wonder lost
Or idle dreams indulging, we obtain
No certainty; but wander far astray
In theories and speculations wild.

To man it was not giv'n to know the whole
Dark mysteries of generations past;
Nor when the potent hand that made the stars,
Did people this small globe with living swarms
Of active moving bodies, gradually
Evolving from each other, thro' the love
Of reproduction and of changes; gifts
Of holy origin, so kindly granted.

Some bounds were set to human scrutiny,
And searching lore. What was and what will be
Often becomes a riddle, else a theme
Too lofty, too obscure and deep. Let us
Apply the soaring intellect to facts;
Let us but try to know, survey, enquire
And prize what is, this study to admire,
Most useful to us all, while here we live.
Beware thou daring man to dive too deep
Into the abyss of eternity,
Before thou was or after death will be.
The present is thy own, the past so far
As memory can reach, the future is
Into the hands of God, who rules the whole
Of time and existence, in endless course.

Thus I shall not attempt to raise the veil
That hides the earthly doom and human fate,
In times to come. I must myself confine
To past and present years, what is displayed
To mortal view, and I delight to study.

Notes on quotations and poems

Edward Hitchcock Jr (1849). The poetry of geology. *The Indicator* **2**, 109–11.

> Edward Hitchcock Jr (1828–1911) was a member of the class of 1849 at Amherst College, Amherst, Massachusetts, where his father, a famous geologist, was President. The younger Edward went on to become Professor of Physical Education and Hygiene at Amherst. His essay, 'The poetry of geology,' appeared in the 1849 volume of the College's short-lived literary periodical *The Indicator* (1848–50).

Anon. (1836). To a fossil fern. *The Museum of Foreign Literature and Science* **29**, 572.

> *The Museum of Foreign Literature and Science* (1822–42) was a popular Philadelphia-based eclectic magazine which contained short stories, poetry, and 'literary and scientific intelligence' excerpted from European sources. This anonymous sonnet was originally published in *The Court Magazine and Monthly Critic*, a London 'ladies magazine.'

Anon. (1850). The coal and the diamond. *Merry's Museum* **20**, 78.

> *Merry's Museum* (1841–72) was one of the most successful popular American literary magazines. Poems appeared monthly in this New York-based journal, and were usually unsigned. The chemical similarity between coal and diamond, as well as the thinly veiled abolitionist sentiments of the author, would have been well understood by most nineteenth-century readers.

John Scafe (1820). *A geological primer in verse: with, a poetical geognosy; or feasting and fighting, and sundry other right pleasant poems with notes.* London: Longman, Hurst, Rees, Orme, and Brown, 61 pp. (Excerpts were republished in *The American Journal of Science* **5**, 272–85.)

> John Scafe was presumably associated with the University of Oxford at the time he wrote his whimsical geological poems which relate the succession and distribution of British strata along with their characteristic minerals and fossils. Scafe acknowledged the advice and encouragement of William Conybeare of Christ's Church, Oxford, and William Buckland, Professor of Geology at the University. Little else is recorded about his geological endeavors, and no subsequent reference to him has been found.
>
> 'The birth of granite,' written in 1811 but not published until 1820, is a description of the mineralogy of granite, which is composed primarily of quartz (silex), feldspar (an alkali aluminum silicate), and an iron–magnesium silicate, commonly mica. Granite was believed to be the basement rock on which all other formations were deposited.
>
> 'Geological cookery' provides recipes for several common rock types.

'Puddingstone' is an archaic term for conglomerate, and 'amygdaloid' refers to basalt with mineralized cavities.

In 'A poetical geognosy' Scafe outlined the stratigraphy of Britain using the Neptunian concepts of Abraham Werner. Werner believed that the world's rocks were deposited in a regular series of formations precipitated from a ubiquitous ocean. Neptune thus invited the rocks to his table in a certain order: first granite, followed by gneiss and mica slates, and so on. A few formations of limited distribution such as the 'calcaire grossier,' a Tertiary marine limestone of the Paris basin, are given special mention. Characteristic fossils contained in each formation are noted as the dishes served to each guest. Disruptions of the rock sequence caused by 'some dislocations and many wry faces' (i.e. facies), 'many faults,' and finally the rages of Pluto (igneous rocks) add a certain confusion to Neptune's once-orderly table.

John Scafe published two other small volumes of geological poetry:

John Scafe (1820). *King Coal's levee, or geological etiquette, with explanatory notes; and the council of the metals. Fourth edition. To which is added Baron Basalt's tour.* London: Longman, Hurst, Rees, Orme and Brown, 119 pp.

John Scafe (1820). *Court News; or, the peers of King Coal, and the errants; or, a survey of British strata, with explanatory notes.* London: Longman, Hurst, Rees, Orme and Brown, 61 pp.

King Coal's levee is an ambitious 1145-line poem 'intended to exhibit geological features, and order of stratification of England and Wales.' 'The council of the metals' and 'Baron Basalt's tour' give British localities of metallic ores and basalt, respectively.

Anon. (1851). On the entrance to the Mammoth Cave. In *Pictorial guide to the Mammoth Cave, Kentucky* by Horace Martin. New York: Stringer & Townsend, 108 pp.

Mammoth Cave, carved from the massive Mississippian limestones of west-central Kentucky, was well known by the 1840s as one of America's greatest natural wonders. Guidebooks to the cavern, such as the 1851 *Pictorial guide* of the Reverend Horace Martin, were popular with tourists. An anonymous visitor wrote 'On the entrance to the Mammoth Cave,' which appeared as the first of several poems in an appendix to Martin's book.

G. D. Prentice (1845). Mammoth Cave. *The Living Age* **7**, 53.

George Dennison Prentice (1802–70), author of 'Mammoth Cave,' was editor of the *Louisville* (Kentucky) *Journal*, a newspaper where the poem first appeared. *The Living Age*, a popular Boston monthly, reprinted the poem, as did several guide books to the cave.

Anon. (1756). *Lines made after the great earthquake, in 1755.* Boston: broadside.

On November 1, 1755, a devastating earthquake destroyed much of Lisbon, Portugal, and the surrounding region. Tens of thousands of people were killed or injured, as the All Saints' Day morning shock trapped worshippers in collapsed churches. Three weeks later an earthquake of far less severity rocked Boston, Massachusetts, and the adjacent New England towns. These dramatic events were inevitably linked, and were the inspiration for scores of earthquake-related religious publications. The anonymous broadside poem, *Lines made after the great earthquake*, appeared as a public exhortation shortly after the New England shocks. The author mistakenly assumed that earthquakes in Europe, North America, and South America were related. (In fact, no other mention of a November 1755 earthquake in South America has been located.) Of the 36 verses originally published, verses 1–6, 8, 21–23, and 34–36 are reproduced here. The remaining 23 stanzas are devoted for the most part to the advantages of repentence and the consequences of sin.

'Flaccus,' pseudonym (1841). Sensations and reflections, caused by the earthquake in January last. *The Knickerbocker; or New-York Monthly Magazine* **18**, 27–28.

On January 26, 1841, a minor earthquake startled residents of New York City and nearby Newark, New Jersey. *The Knickerbocker*, a conservative New York literary and political periodical, published this poem in July of that same year. The sentiments of the anonymous 'Flaccus' regarding God's purposes in the event are similar to those of the Boston broadside poet of 85 years earlier.

James Montgomery (1827). *The pelican island, and other poems.* Philadelphia: B. Littell, 156 pp. (excerpts from Canto II).

James Montgomery (1819). *Greenland, and other poems.* London: Longman, Hurst, Rees, Orme and Brown, 250 pp. (excerpts from Canto II).

The British poet and journalist James Montgomery (1771–1854) created some of the finest geological imagery in verse in his epic poems *The pelican island* and *Greenland*. The description of coral island formation in *The pelican island* was perhaps the most often quoted geological poem of the nineteenth century, for it appeared in dozens of geological textbooks and treatises.

Greenland contains a vivid descriptive account of Icelandic volcanism. The 1783 eruption of the volcano Lakagigar, which produced the largest lava flow in recorded history and killed nearly 20 per cent of Iceland's inhabitants with its poisonous gases and ash, may have been the inspiration for these verses.

Felicia Dorothea Hemans (1836). Epitaph on a mineralogist. *American Magazine of Useful and Entertaining Knowledge* **3**, 72.

Felicia Dorothea Hemans (1793–1835) was a successful British poet whose works were reprinted many times in the nineteenth century. Known as 'the poet of the affections,' she was praised by contemporaries for her 'felicity of expression and delicacy of sentiment,' and her best-known works are characterized by pathos and intensity of emotions. 'Epitaph on a mineralogist' represents a lighter side of Hemans' poetry. The deceased British mineral collector noted in the poem is not identified, though the moral of the poem is unambiguous.

Anon. (1848). 'The miner lad' and 'The miner's doom'. Both in *The coal regions of Pennsylvania* by Eli Bowen. Pottsville, Pennsylvania: B. Carvalic, 72 pp.

The coal mines of Pennsylvania attracted hundreds of British miners in the 1830s and 1840s. Bowen's *Coal regions* was published for the workers, and portions were printed in Welsh. The last few pages were devoted to poetry, which romanticized the laborer and his sufferings. The danger of methane gas explosions in coal mines, as described in 'The miner's doom,' is still a fact of life in this profession.

George Fleming Richardson (1838). *Sketches in prose and verse. Containing visits to the Mantellian Museum, descriptive of that collection.* London: Relfe & Fletcher, 324 pp.

George Fleming Richardson (1796–1848) distinguished himself both as geologist and poet. He was curator of Gideon Algernon Mantell's remarkable private museum at Lewes, south of London, and later became curator at the British Museum, where Mantell's collection was eventually housed. Mantell collected every sort of natural curiosity, though fossils remained his abiding passion. His most famous discoveries were the giant *Iguanodon* and *Hylaeosaurus*, whose bones were excavated from the limestone quarries of Tilgate Forest. The Mantellian Museum, based at the collector's residence, grew to such a size that his wife and children were forced to seek lodgings elsewhere.

Richardson's poems describe three aspects of Mantell's activities. The first poem is a brief introduction to the collection. 'The nautilus and the ammonite' recounts the similarity between two types of chambered coiled cephalopods in Mantell's collection: the ammonite which became extinct after the Mesozoic and the nautilus which still survives in a region of the southwestern Pacific. The third poem, 'A ryghte trewe storie,' relates one of Mantell's popular geological and historical excursions. Only 23 of the original 40 stanzas are reproduced here; the remaining lines are devoted to Mantell's observations on British history.

Anon. (1826). A meditation on Rhode-Island coal. *New York Review or Athenaeum Magazine* **2**, 386–388.

Anthracite was discovered at Portsmouth on Rhode Island, in the state of the same name, in the eighteenth century. Shortly thereafter the Rhode Island Coal Company, incorporated in 1808, became one of the first

American companies to exploit fossil fuels. In spite of the obvious advantage to New Englanders afforded by a local supply of good quality coal, the Rhode Island Company had difficulty convincing many potential buyers of the usefulness of their anthracite. 'A meditation on Rhode-Island coal,' which appeared in *New York Review or Athenaeum Magazine* (a short-lived publication, not to be confused with the successful *New York Review* first issued in 1837), may have been part of the Company's active advertising and publicity campaign.

The author's meditation touches on events of the 1820s as well as coal. The first stanza contains references to Thomas Moore's (1779–1852) popular epic poem *Lalla Rookh*, published in 1817, and the annual report of the United States Department of the Treasury.

Stanzas 10 and 11 probably refer to a ceremony in August (not July) 1824, when the French General La Fayette made a triumphant return visit to the United States, where he had played an important role in the American Revolution. The 'eastern governor' may have been Governor William Eustis, a surgeon during the Revolution and later Secretary of War, who hosted the popular French hero in Boston, Massachusetts. John Adams, a leader of the American war for independence and second President of the United States, attended the festival, which was a short journey from his retirement residence in Quincy, Massachusetts.

'Jonathan' in Stanza 11 may refer to a well-known character in the play *The contrast* (1787) by Royall Tyler. The first of many 'stage Yankees,' Jonathan represented the pragmatism and rural virtues of the common American. 'Bowing to Jonathan' probably signifies Eustis's attempts to acknowledge the common people of Boston.

Reuben Sears (1819). *A poem on the mineral waters of Ballston and Saratoga, with notes illustrating the history of the springs and adjacent country.* Ballston Spa, New York: J. Comstock, 108 pp.

Mineral hot springs were discovered in Saratoga County, New York, shortly before the American Revolution. Their convenient location, about 200 miles north of New York City and 12 miles west of the navigable Hudson River, made the springs at Ballston and Saratoga a mecca for the wealthy seeking relief from discomfort and diseases of the summer months. The principal constituents of the carbonated waters, as related in Reuben Sears's poem, are sodium chloride, calcium carbonate and magnesium carbonate, with lesser amounts of sodium iodide, sodium bicarbonate, and iron carbonate. In spite of Sears's description of the springs as 'the cup of health,' other contemporary writers questioned whether the improved condition of many visitors might not be as much owing to 'change of air, exercise, temperance, and the influence of expectations' (*New England Journal of Medicine and Surgery* (1817) **6**, 363–385).

Anon. (1847). The geologist's wife. *Living Age* **12**, 231.

'The geologist's wife' was written in Britain, though the original source and the poet, 'Susan,' are unidentified by the *Living Age*. The Scottish Highland localities mentioned in the poem would make an arduous two-week field trip.

Constantine Samuel Rafinesque (1836). *The world, or instability, a poem with notes and illustrations.* Philadelphia and London: C. S. Rafinesque, 248 pp.

Constantine Samuel Rafinesque (1783–1840) was a curious figure in the history of American natural sciences. Educated in France and Italy, he first visited the United States from 1802 to 1805, during which time he assembled a botanical collection in Pennsylvania and Delaware. After 10 years of studying and teaching in Palermo, Sicily, he returned to North America to resume his botanical investigations. From 1818 to 1826 he was Professor at Transylvania University at Lexington, Kentucky, and he subsequently moved to Philadelphia where he lectured at the Franklin Institute.

A prolific writer, Rafinesque's bibliography totals almost 1000 titles, including many book-length works. His natural history studies were criticized for indiscriminate introduction of many ill defined species, especially plants. He not infrequently antagonized publishers and readers through his vicious attacks on scientists, politicians, and other well known figures with whom he disagreed. Consequently, Rafinesque became his own publisher for many of his most ambitious projects, such as *The world*.

The world, or instability, an epic poem in 5400 lines, was one of Rafinesque's strangest productions. The work was published under the pseudonym 'Constantine Jobson,' though Rafinesque is listed as the publisher and owner of copyright. The 20 chapters (of which only Ch. 5 and portions of Ch. 4 are reproduced here) encompassed all aspects of natural history, as well as such diverse themes as love, religion, war and peace, women, sin, and 'ultimate prospects of the Earth and mankind.' In the 'Editor's preface,' written by Rafinesque, the poem is praised as comparable with the best works of Pope and Milton and 'superior in some points.' He states 'the great aim of this poem is to prove that *Instability* is as much a law of nature, as attraction or gravitation; [and] that it rules both the physical and moral worlds. . . . It is as if Newton had explained his laws of attraction and repulsion in a poem, instead of a mathematical work.' In conclusion, Rafinesque proclaims that 'those who may dislike this poem must have a bad heart, be exclusive in opinion, or fond of strife and discord.' Rafinesque was not a modest man.

Excerpts entitled 'Mineral springs' and 'Volcanoes' comprise lines 791–814, and 866–908, respectively, of Chapter 4. This chapter, entitled 'The Earth and Moon,' is descriptive of the present state of the oceans, lands, hydrology, and climate of the Earth, and concludes with speculations on the nature of the lakes, rivers, plants and animals that will inevitably be discovered on the Moon.

Chapter 5, 'The former Earth,' is subtitled 'Cataclysms, floods and fossils,' and is central to Rafinesque's theme of 'instability.' Rafinesque's description of the gradual evolution of living forms is, perhaps, a prelude to Darwin:

> 'All is born to live in changeful mood,
> So slowly newer shapes assuming, that
> By mortal eyes, but seldom 'tis perceived.
> What is an age to God? or thousand years?
> He chose they should forever change, and this
> They do by human eyes unseen'

Nevertheless, Rafinesque did not choose to speculate on the mechanism of change, and instead attacked Lamarck for his 'theory so fanciful.' Praise is reserved for Georges Cuvier (1769–1832), eminent French naturalist and founder of the study of comparative anatomy. Cuvier's successful reconstruction of fossil vertebrates is related in this chapter.

Notes on the illustrations

Woodcuts reproduced in this volume are from the following geological texts of the nineteenth century.

G. A. Mantell (1839). *The wonders of geology*, 3rd edn. London: Relfe and Fletcher, 2 vols.

L. C. Beck (1842). *Mineralogy of New-York; comprising detailed descriptions of the minerals hitherto found in the State of New-York, and notices of their uses in the arts and agriculture*. Albany, New York: W. and A. White and J. Visscher, 539 pp.

E. Hitchcock (1845). *Elementary geology*, 3rd edn. New York: Mark H. Newman, 352 pp.

D. Page (1849). *Elements of geology*. New York: A. S. Barnes, 332 pp.

S. St John (1851). *Elements of geology*. New York: G. P. Putnam, 334 pp.

J. Le Conte (1884). *Compend of geology*. New York: Appleton, 398 pp.